The Healthiest City

MILWAUKEE AND THE POLITICS OF
HEALTH REFORM

The Healthiest City

Milwaukee
and the Politics of
Health Reform

JUDITH WALZER LEAVITT

The University of Wisconsin Press

The University of Wisconsin Press
114 North Murray Street
Madison, Wisconsin 53715

3 Henrietta Street
London WC2E 8LU, England

First published in 1982 by Princeton University Press

Library of Congress Cataloging-in-Publication Data
Leavitt, Judith Walzer.
The healthiest city: Milwaukee and the politics of health reform
Judith Walzer Leavitt.
318 pp.   cm.
Includes bibliographical references and index.
ISBN 0-299-15164-6 (pbk.: alk. paper)
1. Public health—Wisconsin—Milwaukee—History. 2. Public health
administration—Wisconsin—Milwaukee—History. 3. Health care
reform—Wisconsin—Milwaukee—History. 4. Urban health—Wisconsin—
Milwaukee—History. 5. Milwaukee (Wis.)—Politics and government.
I. Title.
RA448.M5L4   1996
362.1'09775'95—dc20   96-5117

FOR MY MOTHER AND FATHER

Sally Hochman Walzer
Joseph P. Walzer

Contents

List of Illustrations

List of Tables

Preface

"Death in our drink," warned the Milwaukee news-paper editors advising citizens of danger in their drinking water.[1] This sounds like a quotation from a newspaper during the city's cryptosporidiosis outbreak in 1993, but the year was 1879. Five years earlier, Milwaukee had proudly opened its first municipal water works, and the officials who worked so hard to persuade the city to spend its money on pipes and pumps in the name of public health did not welcome the news of "unwholesome, unhealthy fluid."[2]

Then—as now—bringing clean and safe water to the rapidly growing city was neither easy nor cheap. But during the last decades of the nineteenth century voters let their representatives know that safe drinking water was important to them. Gradually, new water intakes, more efficient pumps, sewage treatment, and water purification paid dividends in better health for the city.

Because of the high cost of such sanitation projects, city politicians needed to be convinced that voters would keep them in office if they supported hefty municipal expenditures. The strong progressive tradition in city politics helped pave the way. Socialist Mayor Daniel Hoan swept

[1] The quotation is from the *Milwaukee Daily News*, August 28, 1879. Sections of this new preface are drawn from my Op Ed article written for the Progressive Media Project and published in the *Milwaukee Sentinel*, April 20, 1993, *Chicago Tribune*, April 16, 1993, among other newspapers. I am grateful to the Project for its permission to use this material here.

[2] *Evening Wisconsin*, July 29, 1879.

xiii

into office in 1910 under the banner of free health care for all Milwaukeeans. This never came to pass, but by 1916 citizens, proud of their clean water, embraced the necessity of preventive public health projects.

Updated and expanded in the 1930s, the safe water system helped earn Milwaukee the title of "the healthiest city" in the 1930s and 1940s. Milwaukee consistently won first or second place in the national health conservation contests jointly sponsored by the U.S. Chamber of Commerce and the American Public Health Association. It won so often because expenditures for illness prevention proved that citizen health was among the city's highest priorities.

This book tells the story of how Milwaukee public health officials and their supporters convinced the city politicians that it was necessary and beneficial to spend public money on preventing illness and promoting citizen health. It underscores the connections between politics and public health and provides a specific historical example of how one city solved some of its most devastating health problems. Focusing on the three main areas of public health activity in the nineteenth and early twentieth centuries— infectious disease control, sanitation and the environment, and food regulations—the book demonstrates that the road to health reform was neither direct nor easy. In the hands of medical officers who often knew their medicine better than their politics, health policy encountered detours and suffered from mistakes. But the outcome of improved health indicators, including decreasing mortality rates and increasing life expectancy, in this midwestern city did prove that positive changes could be accomplished, that citizen health could be measurably improved through public programs and regulations.

Today, at the end of the twentieth century, we are again faced with difficult public health problems—emerging new viruses, the return of old diseases in new drug-resistant forms, new environmental challenges—many of which, again, seem to defy solution. The reissue of this book fourteen years after its first publication indicates a continuing interest in the important lessons that our past experiences can teach. Today's issues are different from those apparent in the late nineteenth century in one important way. Through the course of the twentieth century the federal government has become increasingly involved in public health activity, most significantly through the development of regulations and standards that local communities are now required to meet. The partnership of federal, state, and local governments working together to protect the public's health has proven effective, and the health of most Americans has improved measurably.

Unfortunately, in recent years Milwaukeeans—and Americans generally—have paid less attention than they should to the gears that keep the machinery of our cities running, and the health of the populace has suffered. "Taxes" and "public spending" have become noxious phrases to politicians worried about reelection, and commitment to some of the most basic services has fallen to alarming levels. Political pressures to decrease federal environmental regulations are strong.

The bill for this disinvestment is now coming due in various areas relevant to the public's health. One stark example is that our water supplies are threatened with parasites most of us had never before heard of—for example, cryptosporidia. In 1993, more than 400,000 Milwaukee-area citizens learned the hard way, through bouts of severe gastrointestinal distress, that the city had not adequately monitored and updated its aging water system. The municipal water supply, serving almost one million people, became contaminated with an organism that compromised the health of people who had trustingly turned on their water taps. Although at the time most Milwaukeeans suffered only the inconvenience and discomfort of stomach cramps and diarrhea, for people with weakened immune systems cryptosporidium "infection can be unrelenting and fatal."[3] Following the outbreak forty-three people died.[4] "Death in our drink" is again headline news.

Actually, the Milwaukee water system in April 1993 was operating within existing state and federal criteria for acceptable drinking water quality. There were no regulatory standards for the parasite that caused all the problems. Cryptosporidium, recognized as a human pathogen only since 1976, is resistant to chemical disinfectants and can only with difficulty be physically removed from water through microfiltration. The 1993 Milwaukee experience and outbreaks like it in recent years in communities around the country emphasize the need for continuing study and surveillance by public health agencies that can develop appropriate responses, monitoring techniques, and regulations as new risks arise. The public health infrastructure needs constant attention.[5]

If something positive can come out of Milwaukee's recent suffering—and this can be a lesson for cities across the country—it is the realization that public health and welfare are issues that should concern us all. The cost of not tending to people's basic needs is just too great. It is measured in inconvenience, in illness, and even in death. Rereading public health history and learning from past experience can help keep our eyes open to the dangers of neglect.

The job of government is by constitutional and legislative definition to protect and preserve the public good. Our historical experiences suggest that this means supporting

[3] W. R. MacKenzie, M. S. Hoxie, M. E. Proctor, et al., "A Massive Outbreak in Milwaukee of Cryptosporidium Infection Transmitted through the Public Water Supply, *New England Journal of Medicine* 331 (1994): 161.

[4] The figure is provided by Sara Terry, "Drinking Water Comes to a Boil," *New York Times Magazine,* September 26, 1993, p. 45, based on reports from the Milwaukee AIDS Project staff.

[5] On the Milwaukee outbreak, see MacKenzie, Hoxie, Proctor, et al., "A Massive Outbreak in Milwaukee of Cryptosporidium Infection," pp. 161–67; Terry; "Drinking Water Comes to a Boil," pp. 42–45, 48, 62, 65; "Assessing the Public Health Threat Associated with Waterborne Cryptosporidiosis: Report of a Workshop," *Morbidity and Mortality Weekly Report* 44 (June 16, 1995): 1–16.

urban and state public services, enforcing the clean air and water acts, and encouraging agencies—such as the Environmental Protection Agency—in their health work. Public health and the prevention of disease can be enhanced when we articulate their importance and convince our representatives that we value health and see an important role for government in its promotion. We are so easily upset by the high cost of building the well that we rarely stop to think that we all need to drink from it.

Madison, Wisconsin
December 1995

Acknowledgments

It is a pleasure for me to be able to thank all the people who have helped me while I worked on this book. My study of public health in Milwaukee originated as a doctoral dissertation at the University of Chicago under the supervision of Richard C. Wade, Lester S. King, and Arthur Mann. I am immeasurably grateful to these historians. John Hope Franklin at the University of Chicago also provided important support during my years as a graduate student.

A Maurice L. Richardson fellowship and a faculty grant from the Graduate School of the University of Wisconsin provided financial assistance. This publication also was supported in part by NIH Grant LM 03052 from the National Library of Medicine. Portions of chapters three and four appeared in the *Bulletin of the History of Medicine* (Vol. 50, 1976) and the *Journal of the History of Medicine and Allied Sciences* (Vol. 35, 1980) and are reproduced here with the permission of the editors.

My work could not have proceeded without the cheerful cooperation of the librarians at the State Historical Society of Wisconsin, where I spent many long hours. I especially want to thank Wilma Thompson in the Microforms room and the staff of the Archives division. Myrna Williamson in the Iconography section aided my search for illustrations. Dorothy Whitcomb, historical librarian, and Blanche Singer, interlibrary loans, at the Middleton Health Sciences Library at the University of Wisconsin proved equally helpful. The University of Wisconsin Cartography Laboratory drew the maps. At the City of Milwaukee Health Department Robert J. Harris, Jr. facilitated my access to the official records and photographs. At the Milwaukee Public Library Paul J. Woehrmann assisted my searches through the Local History Room collection and Herbert Rice graciously allowed me to use his working index to the Milwaukee *Sentinel.* The late Mary Dougherty of the Milwaukee Academy of Medicine patiently guided me through the uncatalogued collection.

My colleagues at the University of Wisconsin History of Medicine Department, Guenter B. Risse and William Coleman, generously gave support at crucial moments and challenged my thinking on the major issues in the history of medicine and public health. I am most eager to thank my colleague and friend, Ronald L. Numbers, without whose consistent encouragement and support this book might not have been finished. His high standards of scholarship and prolific pen have provided an important model, and his specific suggestions on this manuscript have been invaluable. I would also like to thank Barbara G. Rosenkrantz, of Harvard University, who read and criticized two chapters; Clay McShane, of Northeastern University, who pushed me over an important hump; Michael L. Walzer, of the Institute for Advanced Study at Princeton, who read an earlier draft of the entire manuscript, and, in a decidedly brotherly manner, offered his suggestions for revisions; and Lewis A. Leavitt, of the University of Wisconsin, whose unrelenting support in ways intellectual, emotional, and practical remains essential to my work.

Others have helped in special ways. Betty Taylor, Piper Luetke, Eve Mokotoff, Christina Ginter, and Deb Goldstein have devoted hours of attention to my children. Sarah Abigail and David Isaac themselves contributed unique brands of support and helped to synthesize my world view. Kathy Conklin and Margaret Kraak patiently typed and retyped portions of the manuscript. William J. Orr, Jr. gathered local election results and translated Polish and German newspaper editorials.

My parents, to whom I dedicate this book, provided the basic ingredients of a supportive and loving family, from which all else grows.

Mistakes, foibles, and inconsistencies remain my own.

Abbreviations

MHD.   Milwaukee Health Department
MMS.   Milwaukee Medical Society
MCMA.   Milwaukee City Medical Association
MAM.   Milwaukee Academy of Medicine
MPL.   Milwaukee Public Library
AJPH.   American Journal of Public Health
WSHS.   Wisconsin State Historical Society
JAMA.   Journal of the American Medical Association

The Milwaukee Health Department Annual Reports have different titles almost every year. To eliminate confusion I have abbreviated each report as MHD, *Annual Report,* year.

The Healthiest City

MILWAUKEE AND THE POLITICS OF

HEALTH REFORM

Under the benignant and guiding influence which you created, you have seen Milwaukee become the first primary wheat market in the world, the fourth pork packing city in the Union, the second commercial city on Lake Michigan, the seventeenth in population, and, according to Dr. Johnson, the healthiest American city.—Dr. Enoch Chase, speaking to the Old Settler's Club, July 4, 1872

It is in health that cities grow; in sunshine that their monuments are builded. It is in disease that they are wrecked; in pestilence that effort ceases and hope dies.—Annual Report of the Commissioner of Health, Milwaukee, 1911

Introduction

During *the last* third of the nineteenth century Milwaukee seemed unlikely ever to earn a national reputation as the healthiest city in America. Rapid population growth and expanding industrialization overwhelmed the city and created an environment characterized by overcrowding, pollution, and high death rates. Milwaukee's population exploded from 20,000 in 1850 to nearly 300,000 by the turn of the twentieth century. Its economy rapidly expanded, its streets bustled with traffic and activity, and its pace of life quickened. Because of its growth and opportunity, Milwaukee offered an attractive and promising home to thousands of immigrants, especially Germans. Along with rapid growth and energizing progress, however, came the negative aspects of urbanization: infectious diseases; crowded, dark, unventilated housing; streets mired in horse manure and littered with refuse; inadequate water supplies; unemptied privy vaults; open sewers; and incredible stench. Many urbanites associated these problems with high urban death rates; thus Milwaukee, along with other expanding American cities, spent considerable effort attempting to relieve its physical affliction. This book focuses on those municipal efforts and analyzes how Milwaukee conquered its vicious health problems and emerged a frequent victor in the national health conservation contests of the twentieth century.

The scope of the municipal government's responsibility

for public health slowly broadened between the incorporation of the city in 1846 and 1910, when the Socialists won the local elections. During those years Milwaukeeans rejected free enterprise as the principle determining governmental action and adopted a clear philosophy of communal responsibility for the health of citizens. The first indication that the city recognized any continuing obligation for public welfare came in 1867, when it established a permanent Board of Health. Slowly thereafter the city health officers became the caretakers of the city's health, and mortality rates began to decline. By 1910 the extent of the government's commitment was evident in the Socialists' winning platform, which promised free medical care for all Milwaukeeans.

The Milwaukee health department focused its activities in three main areas—infectious diseases, sanitation, and food control—each discussed in individual chapters in this book. Control over infectious diseases was the most traditional. Like other American cities, Milwaukee suffered from widespread sickness: smallpox and cholera visited occasionally; diphtheria, scarlet fever, typhoid fever, and tuberculosis were ever present. It was in fighting these infectious diseases that the health department cemented its authority as a legitimate city agency.

Health officers devoted even more energy to sanitation. Many urbanites saw sanitation as the city's most pressing problem, in part because filth was omnipresent and a popular medical theory taught that dirty environments caused disease. In their efforts to improve the health statistics of the city, Milwaukee health officers either directly controlled or monitored garbage collection and disposal, sewage systems, water supply, privies and other nuisances, and industrial pollution.

As part of their battle to improve city health, health officials also found it necessary to control the production and sale of food. Edible commodities, especially milk and meat, carried the potential for transmitting sickness. As the third

Figure 1. Grand Avenue, Milwaukee, 1890s, Courtesy of Milwaukee Public Library.

major part of their attack on unhealthy urban conditions, officials inspected bakeries, confectionaries, slaughter houses, dairies, and ice houses.

Through case studies drawn from each of the three major areas of public health activity in Milwaukee, this book identifies and analyzes the major components of health reform in urban America. Physicians, politicians, and volunteer activists worked together to overcome health problems of great magnitude and to weave the fabric of health reform. Many factors influenced their actions: the medical or technical knowledge available, economic interests, interurban competition, political ideologies, ethnic diversity, corruption, inefficiency, and simple frustration. In order to understand how health reform amalgamated these

5

sometimes conflicting forces and how it grew from piece-meal actions to a comprehensive movement, this study ana-lyzes the process of change in a detailed, local setting. Such single-city studies provide a framework in which to analyze the experience of other cities. From what we already know, it seems that, although the particular events in Milwaukee remain unique to that city, the pattern of health reform that emerges from Milwaukee fits that of other American cities in the late nineteenth century. Municipal bureaucra-cies, although not particularly efficient in solving major health problems, showed themselves capable of conse-quential actions in times of crisis when they had consistent prodding from sanitarians and strong support from influ-ential constituents. During the Progressive period Ameri-can cities conquered the era's most pressing health prob-lems and in so doing accepted responsibility for public health and welfare.

Milwaukee's health reforms were particularly successful because they benefited from widespread community sup-port. At the turn of the twentieth century social reforms that promised to improve urban living conditions were pop-ular rallying issues for an informal alliance between middle-class business reformers and Socialist trade unionists, who put aside ideological differences long enough to defeat the traditional politicians at the polls. A coalition of reform supporters, including upper-middle-class business people, clergy, women's groups, professionals, trade-union Social-ists, Populists, and reform Republicans, united by their dissatisfaction with corrupt municipal regimes, together banished the dishonest officials and corporations that had exploited the rapidly growing economy. This broad-based coalition elected the Socialists in 1910 and supported many of the health reforms that brought Milwaukee out of its nineteenth-century disease-ridden morass to its impressive twentieth-century health achievement. Building on the popular health policies established under the brief Socialist incumbency from 1910 to 1912, the health department continued in the twentieth century to attract wide support. Declining mortality rates portray the beneficial results of the consistent and broad-based municipal health program.

City health activity was not the only determinant of Milwaukee's health, but my study suggests that it was probably the most important during the period from 1860 to 1930. The major killers in the city were respiratory and gastrointestinal infections, and medical science had developed nothing—with the exception of diphtheria antitoxin—to cure the infectious diseases effectively. In contrast, by cleaning up the physical environment and regulating water sources and sewerage, public health reforms significantly reduced exposure to air- and water-borne diseases that thrived in unhygienic, congested urban surroundings. Through preventive isolation, vaccination, and disinfection, public health officials minimized the effects of periodic epidemics. Neighborhood child welfare stations, offering clean milk, feeding advice, and hygiene training, brought significant declines to infant mortality.

This book demonstrates that Milwaukee's mortality declined during the very years that the city waged vigorous sanitation and disease prevention campaigns. The Milwaukee data generally confirm physician Thomas McKeown's observations, based on England and Wales, that curative medicine, before the discoveries in the 1930s and 1940s of the powerful sulphonamides and antibiotics, played only a limited role in improving health statistics.[1] McKeown credits nutrition and public and personal hygiene for improving health and supporting the vast population increases since the eighteenth century. Reversing previous assumptions that advances in medicine had been responsible for in-

---

[1] Thomas McKeown, *The Modern Rise of Population* (New York: Academic Press, 1976) and *The Role of Medicine: Dream, Mirage, or Nemesis?* (Princeton: Princeton University Press, 1979).

creasing life expectancy and decreasing mortality, Mc-
Keown has raised considerable controversy among medical
historians.[2] His thesis gains credibility in the late-nine-
teenth- and early-twentieth-century experiences of Milwaukee.

At the same time that public health activity led to declining mortality rates and offered many Milwaukeeans increased life expectancy, it also put added burdens on people whose lives were already beset by hardships. In the
culturally diverse setting of Milwaukee, health officers met
resistance to their policies, not because people disagreed
with the goals of bringing health to the community, but
because some of the changes contradicted accustomed be-
haviors and added to personal suffering. Thus some individuals refused to stay away from funerals of people who
died from smallpox or refused to allow sick children to go
to the isolation hospital. Others refused to clean up their
cow barns or to placard their homes with infectious disease
signs, or refused vaccination. Because the burdens of the
new policies in terms of cost and pressures to alter life styles

[2] For a flavor of the debate over McKeown's work, see *Health and Society:
The Milbank Memorial Fund Quarterly* 55 (1977): 361-428. Because Mc-
Keown relegates medicine to a minor role in determining human health
over the last three hundred years, he is often linked (incorrectly, I think)
with Ivan Illich, who, in *Medical Nemesis: The Expropriation of Health* (New
York: Pantheon Books, 1976), attacks medical therapy and practice as
being more harmful than beneficial to patients. My book, which is directly
relevant to the issues debated in connection to McKeown's work, does not
address the efficacy or ethicality of medical practice. Statistical analyses
of mortality by American demographers and historians support the importance of public health measures in conquering the important infectious
diseases. See Daniel Scott Smith, "Mortality in the United States before
1900," unpublished typescript, Family and Community History Center,
Newberry Library, 1981; Edward Meeker, "The Improving Health of the
United States, 1850-1915," *Explorations in Economic History* 9 (1972): 353-
373; and Michael R. Haines, "The Use of Model Life Tables to Estimate
Mortality for the United States in the Late Nineteenth Century," *Demography* 16 (1979): 289-312.

seemed to fall disproportionately on the immigrants and the poor, those Milwaukeeans often rejected the methods of reform. This book analyzes the process of urban health reform, positive and negative, accepted and rejected, and seeks to determine how and why municipal governments increased their responsibility for the people's health.

CHAPTER ONE

Milwaukee: The City and Its
Health Problems

From the beginning Milwaukee promised health and prosperity. Situated at the confluence of three rivers and Lake Michigan, the city held what many interpreted to be the key to western settlement: a face to the East and a route to the West. Already an active American Indian trading post when interested speculators arrived in the 1830s, the site seemed ideal for urban commercial growth.[1] It also seemed to be a healthy spot. With the exception of the Menomonee marsh, the area rose in gently rolling hills and enjoyed the benefit of salubrious breezes from the lake. There was every reason for settlers to expect a successful metropolis to emerge from the small village on the western banks of Lake Michigan.

Indeed, Milwaukee fulfilled its promise. In the decades following settlement,[2] it leaped to prominence as a serious

[1] "Thanks to the Indians for choosing the site. The first civilized or semi-civilized people who visited Wisconsin for any but missionary purposes came solely to trade with the aborigines. . . . They went wherever the Indians, with whom it was desired to establish commercial relations, had built their straw-like villages. . . . Their sole object and thought was to find the Indians. They found them at the mouth of the Mahn-ah-waukee river, and there tarried. Hence came 'Milwaukee.' " United States Department of Interior, *Tenth Census of the United States, 1880. Report of the Social Statistics of Cities* 2 (Washington, 1887): 660.

[2] The village was incorporated in 1837, and the eastern and western

rival to other western cities such as Chicago and St. Louis. With an economy based on supplying settlers in the region and on the wheat trade of the upper northwest, the city's population grew. Influxes of German and Irish immigrants joined early English and American residents to provide the base of the new city's population. Many who sought their fortunes in the new Northwest Territory, on its urban frontiers, found what they were looking for in Milwaukee.

Rapid population growth characterized Milwaukee's development in the late nineteenth century. At mid-century, the city housed 20,000 people; by 1880 the population had reached 115,000. During the next decade the population almost doubled to 204,000 and, by 1910, 373,857 people lived in Milwaukee.[3] (See Table 1-1.) This rapid population growth, reflecting immigration, birth rates, and decreasing mortality, provided the labor force for Milwaukee's expanding economy. It also forced the city to reevaluate its responsibilities for citizen welfare. Such rapid expansion

TABLE 1-1
Milwaukee Population Growth

Year	Population	% Increase	Year	Population	% Increase
1850	20,061		1920	457,147	22.3
1860	45,246	125.5	1930	578,249	26.5
1870	71,440	57.9	1940	587,472	15.9
1880	115,587	61.8	1950	637,392	8.5
1890	204,468	76.9	1960	741,324	16.3
1900	285,315	39.5	1970	717,372	− 3.2
1910	373,857	31.0	1980	632,989	− 11.7
				(preliminary)	

parts consolidated in 1839. The city was incorporated in 1846, adding the south side. See Bayrd Still, "The Growth of Milwaukee as Recorded by Contemporaries," *Wisconsin Magazine of History* 21 (1938): 264; and the same author's *Milwaukee: The History of a City* (Madison: The State Historical Society of Wisconsin, 1965).

[3] United States Department of State, *Seventh Census, 1850, Population* (Washington, 1853), p. 922; United States Department of Interior, *Tenth Census, 1880, Statistics of Population* (Washington, 1883), p. 370; *Eleventh Census, 1890, Population* (Washington, 1895), p. 265; Bureau of the Census, *Thirteenth Census, 1910, Population*, vol. 1 (Washington, 1912), p. 670.

soon overwhelmed the churches, private philanthropy, and small city agencies, the traditional caretakers of the sick poor, the homeless, and the jobless. While population itself does not explain the growing responsibility of the municipal government for the welfare of its citizens, wide-scale population increases provided the impetus under which that responsibility developed. The dramatic increases of this period led to Milwaukee's ranking as the twelfth largest city in the United States by 1910.

From its incorporation, Milwaukee has been home to many different peoples. Germans, the major immigrant group to choose the city as their American home, began arriving there in significant numbers in the 1840s. By 1850 more than one-third of Milwaukee's population was German. An economically and politically heterogeneous group, these early German immigrants boasted a high degree of education and skills.[4] German-born immigrants rapidly adapted to life in Milwaukee and became leaders in business and cultural activities. Although they frequently set the pace of life, Germans and their children remained an identifiable and separate group. Not only did most Germans live in close proximity, they also worshipped together, sent their children to German schools, read their own newspapers, and maintained many social organizations exclusively for German-speaking Milwaukeeans.[5]

The Poles formed the second largest immigrant group in late-nineteenth-century Milwaukee. By 1900 six percent of the population in the city were Polish-born and 22 percent were of Polish ancestry.[6] Poles congregated on the

[4] Still, *Milwaukee*, pp. 72, 112-113.

[5] For more on the German population of Milwaukee and its assimilation, see Gerd Korman, *Industrialization, Immigrants and Americanizers: The View from Milwaukee 1866-1921* (Madison: The State Historical Society, 1967); Kathleen Neils Conzen, *Immigrant Milwaukee, 1836-1860: Accommodation and Community in a Frontier City* (Cambridge: Harvard University Press, 1976); Still, *Milwaukee*; and Irve William Zink, "The Influence of the Germans in Milwaukee" (unpublished M.A. thesis, Marquette University, 1941).

[6] In the same year, 19 percent were German-born (54 percent of Ger-

12

south side of the city, mainly in the eleventh and fourteenth wards. The first influx of this group arrived in Milwaukee in the 1870s in response to the availability of unskilled jobs in the rolling mills. The population of the eleventh ward was three-quarters Polish in 1876. In the thirty years after 1880, Milwaukee's south side, particularly its southwest corner, "became almost entirely a Polish community."[7] The Polish immigrants, unskilled and heavily Catholic, never attained the status that the Germans achieved. But because they often held the balance of power in city affairs, politicians could not ignore them.[8]

Other immigrant groups who settled in the city included the Irish, Italians, Russian Jews, Greeks, Bohemians, Hungarians, and Scandinavians. Milwaukee also had a small black community in the nineteenth century. Although these groups were not as visible as the German and Polish immigrants, their presence added to the diversity and cosmopolitan life of the city. In 1890 Milwaukee had the largest percentage of immigrants of any American city.[9]

People born of English-speaking parents, although a minority in Milwaukee, were frequently among the most

---

man descent), 31 percent of the city's total population were foreign-born. For this information, see the United States Census for these years, adapted by Still, *Milwaukee*, pp. 574-575, and Donald Pienkos, "Politics, Religion, and Change in Polish Milwaukee, 1900-1930," *Wisconsin Magazine of History* 61 (1978): 179.

[7] Roger Daniel Simon, "The Expansion of an Industrial City: Milwaukee, 1880-1910" (unpublished Ph.D. dissertation, University of Wisconsin, Madison, 1971), p. 216; MHD, *Annual Report*, 1876, p. 57. Korman noted that by 1905, "the South Side was predominantly Polish" (p. 47). The census data indicates, however, that while the predominant group in the area may have been Polish, many Germans, Scandinavians, and Irish lived among them. See the *Wisconsin State Census Report*, 1895, Parts 1-2 (Madison: State Printer, 1895), p. 88 and *Eleventh United States Census*, 1890, pp. 262-265.

[8] See, for example, during the 1892 local election when the Democrats won with help from a block of Polish votes. *Sentinel*, April 6, 7, 9, 10, 1892.

[9] *Eleventh United States Census, 1890, Population* Part 2, p. lxv. In 1890, one-fifth of Milwaukee's population could not speak English.

socially and economically advantaged. Native-born sons, very influential in city affairs, were "disproportionately represented in the business and managerial occupations."[10]

Milwaukee grew in three distinct geographic sections. The Milwaukee River, running south through the city, formed a peninsula of land known as Juneautown after Solomon Juneau, the original founder. Across the Milwaukee River to the west was Kilbourntown, developed by Byron Kilbourn, another speculator, in direct competition to Juneau's settlement. Although legally bound together, the east and west sides remained distinctive in many respects. Because of the early rivalry, Kilbourn refused to align east-west streets with Juneau's, and both settlements used different street names. Bridges spanning the river today must cross diagonally to match the misaligned streets, a modern vestige of the founders' competitive spirit. The third section of the city, named Walker's Point, after founder George H. Walker, was always the poor cousin, having been settled later on less desirable land south of the Menomonee River. The three sections incorporated as the City of Milwaukee in 1846. (See Fig. 2.) Physical expansion characterized Milwaukee's development from a commercial town to a large industrial center. The 1910 boundaries were significantly larger than the 1846 city limits, and by 1960 the city had swallowed up its original perimeters. (See Fig. 3.)

Milwaukee's most privileged groups lived on the exclusive east side, where large homes along grand boulevards flourished. The southern end contained the business district, surrounded by the crowded homes of the poorer people who served the rich. Most of the east side contained single-family dwellings, giving it the lowest population density of any area of the city. Those who had the most choice about their residence lived along the lake bluffs close to the central city. As Milwaukee expanded, property owners in this section continued building their large homes along

[10] Simon, pp. 269, 82-83.

14

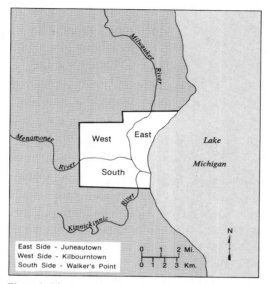

Figure 2. Three sections of Milwaukee, 1846. Courtesy
of University of Wisconsin Cartography Laboratory.

the lake shore, extending the east side farther north. Roger
Simon, who has studied the residential patterns in Milwaukee at the turn of the twentieth century, concludes that the
physical environment of the east side was "vastly superior"
to that of other sections of the city. It had the largest space,
the easiest access to the business district, and the earliest
provision of urban services.[11]

The west side of Milwaukee also attracted some wealthy
people who built large homes along Grand Avenue, but
for the most part this section contained middle-class German immigrants and their families. Parts of the west side
closest to the rivers, like their counterparts on the east side,
housed new immigrants and laborers who lived in crowded
conditions. However, single-family dwellings characterized
the west side. With a population density only slightly higher

[11] Simon, p. 284.

15

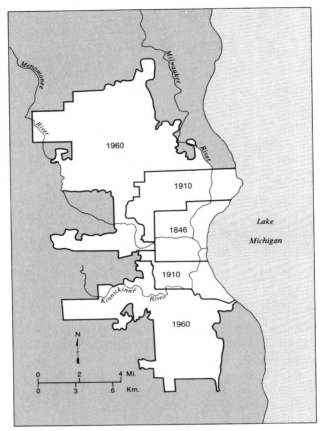

Figure 3. Physical expansion of Milwaukee, 1846, 1910,
1960. Courtesy of University of Wisconsin Cartography Laboratory.

than the east side's, the west side was a desirable place to
live.[12]

The south side was the poorest and least desirable of the
three sections of Milwaukee. In the decades after 1870

[12] United States Department of Interior, *Eleventh United States Census,*
1890, *Vital and Social Statistics,* pp. 258-267; Simon, p. 299; and Rudolph

16

unskilled immigrants, largely Polish, but also German and Bohemian, settled in the south-side wards near their places of employment. These laboring families lived in a markedly inferior environment compared to the rest of the city. Houses were small and crowded, although physically they resembled a middle-class neighborhood more than a New York slum of the same period.[13]

The physical division of Milwaukee into three parts containing varying economic, ethnic, and political interests added to urban tensions in the nineteenth century. These sectional differences led at least one observer to complain that the city, like Gaul, was divided into three parts, resulting in a "triangular duel" that made progress difficult. "The wonder is that, despite this division, topographical and social, the city has grown and thrived."[14]

If beer made Milwaukee famous, commerce in grain initially made the city grow. The rich commercial hinterland and railroad connections provided the stable base on which city development depended in the early years.[15] In addition to commerce, industry also burgeoned. Milwaukee derived its nickname—the Cream City—from its early manufacture of cream-colored bricks produced from local clay. Flour mills, breweries, slaughter houses, leather works, metal trades, cigar works, clothing factories, and iron works added to Milwaukee's economic base. By the turn of the twentieth

---

Hering, "Report on Garbage Disposal by Rudolph Hering to the Mayor and Common Council of the City of Milwaukee," December 23, 1907, reprinted in the Milwaukee Health Department, *Annual Report*, 1908, pp. 155-156. (Hereafter cited as MHD, *Annual Report*, year, despite title variations.)

[13] The pattern of settlement whereby a family purchased a small house and let out rooms to help to pay the mortgage was familiar in urban outskirts in the late nineteenth century. The process in Milwaukee's south side is described by Simon, pp. 216-227. See also, Sam B. Warner, *Streetcar Suburbs: The Process of Growth in Boston 1870-1900* (Cambridge: Harvard University Press, 1962).

[14] Charles King, "The Cream City," *Cosmopolitan* 10 (1891): 554.

[15] Still, *Milwaukee*, pp. 168-178.

century, Milwaukee maintained its commercial activities but relied more heavily on manufacturing.[16]

Industry's dependence on water for power led to decentralized settlement. First locating south along the lake and then west along the Menomonee River, factories sprang up in outlying areas of the city. Some industries removed themselves entirely from city property and established plants to the south and west of the city limits. The resulting occupational polarization that outlying industry encouraged merely accentuated an already existing occupational segregation of Milwaukeeans. Professional groups and clerical workers lived on the east side or directly west of the business district on the west side. Skilled artisans lived in the northwest sections of the city, while those with specific skills resided near their particular industry. Unskilled workers clustered in industrial areas on the south side and at the periphery of the city.[17]

Milwaukee's deep commitment to industry created a large laboring population, which, with the relatively small numbers of industrial proprietors and managers, led to significant economic division in the city. In combination with ethnic divisions, these income and occupational polarities help to explain the political environment in Milwaukee at the end of the nineteenth and the beginning of the twentieth centuries.

Republicans, supported by middle-class and increasingly German voters, dominated post-Civil War politics in Milwaukee, as they did throughout Wisconsin, until two events

[16] Milwaukee's six leading industries in 1909, as identified by Simon, were: (1) tanned, curried and finished leather; (2) malt liquors; (3) foundry and machine shop products; (4) slaughtering and packing; (5) car and general shop construction by steam railroad companies; and (6) iron and steel works and rolling mills (p. 68).

[17] Simon devised a series of maps that illustrate the "index of disproportion" of the various occupational groups in the various wards of the city. See his dissertation, pp. 131-159. See also Clay McShane, *Technology and Reform: Street Railways and The Growth of Milwaukee 1887-1900* (Madison, State Historical Society of Wisconsin, 1974), pp. 84-105.

broke old patterns and left the field open for competition. The 1889 passage of the Bennett Law, which abolished the use of the German language in Wisconsin schools, led to wholesale German defection to the Democrats, who until then had received support largely from Irish and Polish voters. The depression of 1893-1897, which hit Milwaukee severely, provided new challenges to local government and made old political loyalties obsolete. The reshuffling of allegiances revealed the growing political strength of labor and gave impetus for increased governmental action to solve the problems of urban life.[18]

Labor's strength, although divided and unable to prove itself at the polls in the 1880s, pushed the traditional political parties into increasing governmental involvement in citizen welfare. The Municipal League, reflecting middle-class reform sentiment after 1890, aided this movement by emphasizing civil-service reform and responsible city government. Increasing urban problems, especially during the depression, and wide-scale corruption in both public and private sectors helped to convince many other Milwaukeeans of the necessity for local political changes. A particularly unresponsive Democratic regime at the turn of the century led the discontented to support an emerging reform coalition, which continued to grow in the first decade of the twentieth century.

The major groups in the emerging coalition, the trade-union Socialists and the efficiency-oriented middle-class business reformers, disagreed on some fundamental issues. The middle class supported structural changes in city government that would consolidate power in the hands of the

[18] Still, *Milwaukee*, pp. 279-320; Roger E. Wyman, "Voting Behavior in the Progressive Era: Wisconsin as a Case Study," unpublished Ph.D. dissertation, University of Wisconsin, 1970; Thomas Gavett, "The Development of the Labor Movement in Milwaukee," unpublished Ph.D. dissertation, University of Wisconsin, 1957; Robert J. Ulrich, "The Bennett Law of 1889: Education and Politics in Wisconsin," Ph.D. dissertation, University of Wisconsin, 1965.

economically prominent, leading them to seek, for example, city-wide aldermanic elections in their attempt to break up the local power bases. The Socialists derived their strength at the neighborhood or ward level in the working-class and ethnic areas and thus resisted these centralizing efforts. The issue of municipal ownership of the major urban service enterprises (street cars, gas, electricity) separated the two groups even more strongly, the middle-class groups advocating free enterprise in all areas of the economy. However, both groups found that they could equally support social reforms such as public health that sought to ameliorate adverse living conditions throughout the metropolitan area, the Socialists as part of their attempts to improve the lives of the poor, and the business groups because of their concern for city economic health.[19]

[19] Frederick I. Olson, "Milwaukee's Socialist Mayors: End of an Era and Its Beginning," *Historical Messenger*, 16 (March 1960): 3-8. See more particularly, Olson's Ph.D. dissertation, "The Milwaukee Socialists 1897-1941" (unpublished Ph.D. dissertation, Harvard University, 1952); and his "The Socialist Party and the Union in Milwaukee 1900-1912," *Wisconsin Magazine of History* 44 (1961): 110-116. See also Marvin Wachman, *History of the Social Democratic Party of Milwaukee 1897-1910* (Urbana: University of Illinois Press, 1945); David G.Ondercin, "Corruption, Conspiracy and Reform in Milwaukee 1901-1909," *Historical Messenger*, 26 (December 1970): 112-123; and Sally Miller, "Milwaukee: of Ethnicity and Labor," in Bruce M. Stave, ed., *Socialism and the Cities* (Port Washington, N.Y.: National University Publications, 1975), pp. 41-71. For a muckraker's account of Wisconsin politics, see Lincoln Steffens, "Enemies of the Republic," *McClures Magazine* 33 (1904): 577-593.

The distinction between structural reform and social reform and the development of coalition "progressivism," both of which apply to the Milwaukee experience, are developed in James Weinstein, *The Corporate Ideal in the Liberal State: 1900-1918* (Boston: Beacon Press, 1968); Samuel P. Hays, "The Politics of Reform in Municipal Government in the Progressive Era," *Pacific Northwest Quarterly* 55 (1964): 157-169; John D. Buenker, *Urban Liberalism and Progressive Reform* (New York: W. W. Norton & Company, 1978); and Melvin G. Holli, *Reform in Detroit: Hazen S. Pingree and Urban Politics* (New York: Oxford University Press, 1969). As the LaFollette Progressives gained strength in state politics in Wisconsin, the Socialists, advocating many of the same issues, gained strength in Milwaukee.

United mainly by their common abhorrence of governmental corruption and mismanagement, the two groups
joined hands to defeat the traditional entrenched politicians. Such diverse groups as business people, professionals, trade-union Socialists, Populists, and reform Republicans came together in 1910 behind the Socialist banner
because conciliatory leadership in that party promised honesty and efficiency. The 1910 Socialist platform offered
progressive good government reforms that could be supported by unionists and business interests alike. The success
of Social Democracy, attributable more to the urge for
municipal reform than to a devotion to Socialism, more to
a negative response to the old than to a positive feeling for
the new, cemented the growing commitment to municipal
responsibility for community welfare in Milwaukee.[20]

The reform alliance weakened during the Socialist incumbency from 1910 to 1912 because the Socialists, while
keeping to their pledges of honesty and efficiency, also tried
to implement some of their more ideological desires for
municipal ownership and decentralized government. Events
in the twentieth century reveal that, while the Socialists
were temporarily successful in leading the middle-class reform movement, they were not strong enough to maintain
control. The business-oriented "nonpartisan" reformers,
epitomized by Gerhard Bading, who defeated Socialist mayor
Emil Seidel in his bid for reelection in 1912, ultimately
prevailed in Milwaukee, even though the Socialists retained
control of the mayoralty for most of the next fifty years.
Governmental efficiency and economic liberalism marked
the prevailing reform sentiment in Milwaukee, as they did

[20] This interpretation makes particular sense in terms of the middle-
class progressive coalition that elected Emil Seidel in 1910. The Socialist
administration elected was far from purist. Its leader, Victor Berger, was
at all times a political realist, and his pragmatism led to many compromises
with true Socialism as well as to success at the polls. See David Thelen,
The New Citizenship: The Origins of Progressivism in Wisconsin 1885-1900
(Columbia: University of Missouri Press, 1972) for a similar analysis of
Milwaukee politics in the 1890s.

21

in most American cities. Despite the Socialist defeat in 1912, health reform, which Milwaukeeans of all political persuasions believed necessary for urban prosperity, maintained city-wide support. One accomplishment, therefore, of the short-lived political coalitions of the early twentieth century was a popular health reform movement and an increasingly improving health record.

Like other American cities in the nineteenth century, Milwaukee faced a multitude of health problems. Rising mortality rates warned that life in the cities magnified health troubles.[21] Urban life seemed to exacerbate health problems. Infectious diseases spread rapidly through crowded urban areas, whereas in more isolated regions similar diseases remained contained. Cities with commercial economies invited new diseases with each vessel or train arriving from other cities. The urban environment fostered the spread of diseases with crowded, dark, unventilated housing; unpaved streets mired in horse manure and littered with refuse; inadequate or nonexisting water supplies; privy vaults unemptied from one year to the next; stagnant pools of water; ill-functioning open sewers; stench beyond the twentieth-century imagination; and noises from clacking horse hooves, wooden wagon wheels, street railways, and unmuffled industrial machinery. Industrialization and the population explosion accentuated these negative aspects of the urban environment and allowed disease and death to flourish.[22]

[21] Frederick L. Hoffman, "American Mortality Progress During the Last Half Century," in M. P. Ravenel, *A Half Century of Public Health* (New York, 1921), pp. 94-117, discusses urban mortality rates in the nineteenth century. See also his "The General Death-rate of Large American Cities 1871-1904," *Quarterly Publications of the American Statistical Association* 10 (1906-1907): 1-75. See also Daniel Scott Smith, "Mortality in the United States," and Edward Meeker, "The Improving Health of the United States."

[22] For descriptions of the nineteenth-century urban environment, see Lawrence H. Larsen, "Nineteenth Century Street Sanitation: A Study of Filth and Frustration," *Wisconsin Magazine of History* 52 (1968): 239-247; Lawrence H. Larsen, *The Urban West at the End of the Frontier* (Lawrence,

Milwaukee housing exemplified the unhealthy situation. Immigrants occupied squalid, damp rooms in small houses, where in the last decades of the nineteenth century whole families crowded into the limited space and frequently slept in shifts with boarders. Although tenement conditions did not match New York's in scale or density, the typical frame cottage on the southwest side of Milwaukee and elsewhere in the city resembled "a stable where the family herds together like cattle."[23] Not surprisingly, disease thrived in these settings.

Contemporary Milwaukee observers understood the adverse effects on people's health of overcrowding and basement living. The medical community felt helpless to stem the tide of disaster when epidemics spread through densely populated wards. Not only was isolation impossible, but proper home care in terms of ventilation, cleanliness, and diet seemed equally elusive.[24] One physician described the futility of treating diphtheria in the poorer sections of the city in 1893: "We might as well confess it right straight off, that we just have to give up. We have to give up because of the local surroundings over and over again."[25]

Housing represented only part of the problem. A stroll

Kansas: The Regents Press of Kansas, 1978); Joel A. Tarr, "Urban Pollution-Many Long Years Ago," *American Heritage* 22 (1971): 65-69 + ; Ann Cook, Marilyn Gittell, and Herb Mack, eds., *City Life 1865-1900: View of Urban America* (New York: Praeger Publishers, 1973); and appropriate passages of various urban biographies and histories, for example, Blake McKelvey, *The Urbanization of America* (New York, 1963); Sam Bass Warner, Jr., *The Urban Wilderness* (New York: Harper & Row, 1972); David Ward, *Cities and Immigrants: A Geography of Change in Nineteenth Century America* (New York: Oxford University Press, 1971); Still, *Milwaukee.*

[23] *Sentinel*, March 19, 1904, quoted in Still, *Milwaukee*, p. 390n. See also Carl D. Thompson, "The Housing Awakening: Socialists and Slums, Milwaukee," *The Survey* 25 (1910): 367-376.

[24] Bureau of Labor and Industrial Statistics, Wisconsin, *Twelfth Biennial Report*, 1905-1906 (Madison: State Printer, 1906), pp. 269-354.

[25] Dr. A. B. Farnham during a discussion at the meeting of the Milwaukee Medical Society, April 25, 1893. Minutes in the Milwaukee Academy of Medicine.

down urban streets was not only offensive to the eyes, nose, and ears, but also a threat to health. Tons of horse manure covered the thoroughfares. On rainy days, when the excrement mixed with mud from unpaved streets, pedestrians unavoidably picked it up on shoes and long skirts. Urbanites left garbage in the streets, and dead horses lay for days, if not weeks, where they had fallen. Traffic on urban streets, including herds of animals on the way to slaughter and cows on the way to pasture, was heavy and hazardous; noise levels, deafening.

Near the factories and rendering plants people encountered yet another urban offense: the stench. When a gentle breeze blew across the city, "[d]ecayed animal matter commingled with the aroma from the marsh, forming a perfume which for *strength* if not for *sweetness*, could not be excelled even in Chicago, the famed city of smells."[26] Privy vaults added to urban smells, while their contents permeated the ground to contaminate the water supply in nearby wells. Unflushed open sewers compounded the odor and unsightliness of Milwaukee streets, frequently made impassable by all the refuse, garbage, manure, and dead animals. Although medical and lay people understood the health hazards posed by the environment—"The city to be healthy must be kept clean," warned an early newspaper editorial[27]—Milwaukeeans suffered under these burdens of industrialization and urbanization for many years. During that time the city's death statistics ranked among the worst in the nation.[28]

Even though mortality statistics give only a partial view of a city's health, historians and demographers typically use death rates to determine a city's degree of healthiness.[29]

[26] *Sentinel*, October 17, 1866, p. 1.

[27] *Ibid.*, June 26, 1851, p. 2.

[28] In 1870 Milwaukee's death rate of 23 per 1,000 population was exceeded only in New Orleans, Nashville, Charleston, and New York. Chicago and Milwaukee tied, and cities like Philadelphia, Buffalo, Detroit, and St. Louis boasted superior records. MHD, *Annual Report*, 1871, p. 24.

[29] The national health conservation contests ranked cities on sanitary

Milwaukee's death rate decreased after the 1880s, suggesting a growing healthfulness in the city. (See Fig. 4.) Mortality data, however, must be further analyzed to uncover the complexities of the reciprocal relationship between disease and urban conditions.

The limitations of such an analysis for nineteenth-century data are significant. Disease identification was, at best, precarious for most medical practitioners. Unschooled in sophisticated diagnosis and unfamiliar with the benefits of laboratory science, physicians relied on symptoms and previous experience to make their diagnoses. The results were bound to encompass large variations. Physicians might have labeled respiratory problems as tuberculosis (also called phthsis and consumption), inflammation of the lungs, congestion of the lungs, pneumonia, croup, bronchitis, or

Figure 4. Mortality Rate Per Thousand Population, Milwaukee, 1870-1915. *Sources*: Milwaukee Health Department Annual Reports and Division of Vital Statistics Records.

---

measures, disease prevention, health promotion, and financial support for health work, in addition to death rates. William Butterworth, "Inter-Chamber Health Conservation Contest," *American Journal of Public Health* 20 (1930): 633-636; George A. Dundon, "Health Conservation," MHD, *Bulletin* 20 (1931): 9-11.

even diphtheria. Furthermore, disease reporting remained spotty. While the numbers of dead may be fairly accurate, reporting of specific disease cases and deaths for the nineteenth and early twentieth centuries is probably not a reflection of the actual situation. Despite the obvious and perhaps severe limitations of health and disease statistics, we can tentatively analyze disease patterns in the urban environment by using the best available data.[30]

Tuberculosis killed more Milwaukeeans, year after year, than any other single disease. (See Table 1-2.) Its bacteria spread through bodily secretions, particularly those expelled while coughing. Crowding, shared drinking and eating utensils, and unsanitary facilities fostered the dissemination of tuberculosis. Although some writers have romanticized its symptoms—loss of weight, energy, and appetite, accompanied by a chronic and bloody cough— victims have not been among those who glorified the disease. Each year tuberculosis caused ten to fourteen percent of Milwaukee's deaths.[31]

[30] Milwaukee health statistics suffer from the same limitations evident in other cities. Not only did physicians vary in their diagnoses, many simply did not report some diseases at all. They became more cooperative about disease reporting as the century closed, and later statistics are more complete than earlier ones. Another problem existed within the health office. Because of limited staff, hand tabulations of disease tolls led to errors and inconsistency between sources. These latter discrepancies tend to be minor. Since the health department issued no annual reports during the 1880s, comparisons among the various diseases by ward encounter yet another hurdle. Realizing the limitations of the data, and the limited base for comparative purposes, I have proceeded with this analysis, hoping to begin a process that can be extended in other cities: the ward-by-ward analysis of important causes of death. I have checked all differing versions of confusing data with the Division of Vital Statistics, Milwaukee Health Department.

[31] I would like to thank Drs. Lester S. King, Guenter B. Risse, and Lewis A. Leavitt for helping me to understand the causes, symptoms, course, and spread of infectious diseases. The case rates and mortality rates discussed on these pages are all derived from Milwaukee Health Department Annual Reports and Monthly Reports from the appropriate years, unless otherwise noted.

Diarrheal diseases represented another major threat to life in Milwaukee. A catchall category of intestinal illnesses chiefly of infancy, it embraced cholera infantum, "summer complaint," dysentery, and other diseases whose common symptom was diarrhea. The diseases spread through bowel excreta under conditions of inadequate sanitary facilities or through contaminated water and milk. Urban life encouraged the dissemination of diarrheal diseases, which, as the statistics indicate, were endemic in Milwaukee.

Diphtheria, with alarming symptoms and mortality rates as high as forty-four percent of cases, was among the most feared diseases of childhood. Death could follow onset within a few days, during which time the victim suffered from an extreme sore throat, fever, and difficult breathing. This terrifying childhood disease spread through oral and nasal secretions under crowded urban conditions or through shared drinking cups in the home or school.

"Convulsions," now regarded as a symptom rather than a disease, were another major cause of death in Milwaukee. This term describes jerking and other involuntary movements of the body brought on by irritation of the cortex of the brain. The underlying diseases might have been meningitis, high blood pressure, kidney disease, brain tumors, trauma to the head, tetanus, or encephalitis. The fact that physicians attributed death to convulsions so often is unfortunate for the historian, since that category hides information that would have been useful for our understanding of nineteenth-century disease patterns.

Respiratory infections, such as pneumonia and bronchitis, characterized by fever, cough, fatigue, shortness of breath, and chest pain, also killed many citizens, as did typhoid fever and scarlet fever, both easily spread in the crowded environment. During Milwaukee's early years malaria was probably endemic, but it had virtually disappeared by the 1860s.[32]

---

[32] Erwin H. Ackerknecht, *Malaria in the Upper Mississippi Valley* (Baltimore: Johns Hopkins Press, 1945), pp. 45-48.

TABLE 1-2

Milwaukee Mortality Statistics

Year	Death Rate	% TB	% Diarrheal	% Diphtheria
1870	23.2	9.9	14.3	6.9
1872	26.6	7.7	18.8	1.2
1874	23.8	8.5	18.5	2.2
1876	23.9	7.6	10.1	4.7
1878	16.9	11.7	8.9	9.8
1880	20.7	11.0	7.5	7.3
1882	20.9	14.1	9.5	4.2
1884	18.8	9.3	14.1	4.2
1886	18.3	10.3	10.3	7.3
1888	18.5	10.1	8.2	3.5
1890	18.4	10.1	6.7	6.4
1892	20.3	9.3	11.5	8.9
1894	17.6	11.3	12.5	5.6
1896	15.0	11.4	12.0	5.1
1898	11.9	12.9	15.4	1.8
1900	14.1	11.6	12.2	3.1
1902	12.7	10.4	10.2	2.2
1904	12.9	11.7	7.9	1.4
1906	13.4	10.2	9.1	1.4
1908	12.4	9.8	9.2	1.6
1910	13.9	9.0	8.4	2.6
1912	13.5	7.0	8.5	2.0
1914	12.1	7.2	4.9	2.7
1916	12.9	5.9	5.5	1.3
1918	14.8	6.0	4.9	.8
1920	11.6	6.0	4.3	2.2

Although these endemic diseases regularly accounted for as much as sixty percent of Milwaukee's mortality, sporadic epidemics caused far greater consternation among Milwaukeeans. Smallpox, for example, occasionally struck the city, causing major disruptions and generating panic among the residents. This response resulted not only from the disease's high mortality, but also from its strangeness and its tendency to leave victims marked and disfigured. Smallpox attacked all ages and, exacerbated by unsanitary con-

TABLE 1-2 (cont.)

% Convulsions	% Pneumonia & Bronchitis	% Scarlet Fever	% Typhoid Fever	% Smallpox
15.6	4.9	3.1	2.9	0
15.3	3.7	.5	2.4	10.3
18.0	5.3	2.5	1.7	.1
17.2	5.9	5.2	1.2	7.3
14.3	7.3	1.5	.01	.7
9.2	4.9	4.2	.0	0
11.6	9.8	2.0	1.3	.4
10.2	6.1	2.3	1.8	0
9.5	7.4	.01	1.6	0
10.5	8.6	1.6	1.8	0
9.0	10.7	.01	1.9	0
6.0	10.2	1.7	1.7	0
5.9	9.7	.0	1.6	5.7
6.3	10.9	.0	1.2	0
3.8	13.8	.0	1.4	0
3.5	13.8	1.7	1.5	0
2.5	16.2	.0	1.3	.02
3.8	15.5	.01	1.0	.02
3.9	15.4	.0	2.1	.02
3.4	13.0	.0	1.2	0
3.1	13.4	1.5	3.3	0
2.6	14.6	1.1	1.9	0
2.6	10.4	1.6	.7	0
1.5	13.1	.7	1.2	0
.8	24.8	.8	.4	.03
1.0	17.2	.5	.2	0

ditions and overcrowding, spread quickly through a city. Because of the fear it generated, smallpox paved the way for many of Milwaukee's health reforms.

Many of the diseases described above affected the younger age groups most heavily. In 1871 children under five years of age accounted for 61 percent of the deaths in Milwaukee. For years this age group accounted for 50 percent or more of Milwaukee's mortality, but by the early twentieth century, this figure had dropped to 35 percent. In 1900 only

29

12 percent of Milwaukee's population was under five, indicating the extent of their disproportionate representation in the mortality tables.[33]

Most physicians blamed the unhealthy urban environment with its "miasmatic poison" for Milwaukee's high mortality. One health officer lamented: "This slaughter of innocents is found, chiefly, in crowded parts of the city, where families are massed together, in filthy, dark, ill-ventilated tenements, surrounded by dirty yards and alleys, foul privies, and imperfect drainage."[34] Analysis of disease and ward data reveals that physicians were at least partially correct in attributing high mortality to the urban environment. A person living in a low-density residential neighborhood and whose parents had been born in America had the best chance of surviving in Milwaukee. Conversely, a person living in a congested neighborhood, born of German or Polish parentage, had the least chance of living a long life.

The following discussion investigates the relationship between death, urban conditions, and ethnicity. It does not totally explain the contours of sickness and health in American cities, but suggests some factors that influenced disease patterns. The nature of the data is such that conclusions must remain tentative. The statistical analysis centers on 1890 and 1900 because mortality data by disease and ward are most complete for those census years.

The highest correlates of general death rates in both 1890 and 1900 were density within dwelling and foreign nativity.[35] (See Table 1-3.) The more congested the housing

[33] 12.3 percent of Milwaukee's population was in the 0 through 4 age group in 1900; 10.1 percent in 1910. H. Yuan Tien, ed., *Milwaukee Metropolitan Area Fact Book, 1940, 1950, 1960* (Madison: University of Wisconsin Press, 1962), p. 23.

[34] M. P. Jewett, President of the Board of Health, in the MHD, *Annual Report*, 1872, p. 7.

[35] The correlation coefficient is the Pearson product moment correlation r. The data for all correlations in the following discussion were collected from the Milwaukee Health Department Annual Reports and the

TABLE 1-3[a]

Total Death Rate

	1890	1900
Density-within-dwelling	+.65[b]	+.51[c]
% Foreign-born	+.43	+.36

[a] This table and subsequent tables were computed using Pearson's product moment correlation r from data collected from Health Department vital statistics and the U.S. Census Reports. See note 35.
[b] p<.01.   [c] p<.02.

and the larger the percentage of foreign-born in a given ward, the greater the chance of high mortality rates. Unfortunately, reliance on ward-based data masks any variations within wards. The correlations were stronger in 1890 than they were in 1900, indicating that by the latter year other factors, perhaps medical services, sanitary improvements, or standards of living, became increasingly important in explaining mortality.

The picture becomes more complicated when we evaluate the pattern of specific diseases within the city. For example, a strong positive correlation connects diarrheal deaths with the foreign-born population in 1890 and 1900. But if we sift out German-born from the rest of foreign-born, no relationship exists between Germans and diarrhea in either year. (See Table 1-4.) One possible explanation of the differences between Germans and other immigrants might be that better German nutrition and hygiene determined this ethnic difference, a suggestion supported by the 1900 age component of the diarrheal death rates. The correlation between diarrhea and under-age-one mortality was not statistically significant. Diarrhea, however, did correlate

U.S. Census, 1890, 1900. The 1890 correlations proved significant at p<.01 r = .589, p<.05 r = .467. The 1900 proved significant at p<.01 r = .549, p<.02 r = .503, p<.05 r = .433. I would like to thank Lewis A. Leavitt and James N. VerHoeve for their help with these computations.

TABLE 1-4
Diarrheal Death Rates

	1890	1900[a]
% Foreign-born	+.636[b]	+.622[b]
% German-born	−.005	−.011
Density-within-dwelling	+.406	+.533
Deaths <1	n.a.	−.187
Deaths <5	n.a.	+.752[b]

[a] Diarrheal death rates computed from 1899 data since information not available on a ward basis for 1900.
[b] $p<.01$.

significantly with the under-age-five deaths, perhaps indicating that once babies were weaned onto cow's milk their chances for infection increased. Thus differing dietary and sanitary habits between Germans and other immigrants may explain the wide discrepancy in their children's diarrheal deaths.

The differences between German-born and other foreign-born might also reflect the geographic patterns of immigrant settlement in Milwaukee. We have already noted that Germans came to the city early and settled on the west side in a middle-class, noncongested area. Later immigrant groups without skills or money settled in the more densely populated portions of the central city or the south side.[36] We know that congestion influenced mortality, and it seems logical to assume that lack of sanitary facilities in an overcrowded environment fostered diarrhea. But density does not correlate with diarrheal deaths as strongly as it does with general death rates. Geographer David Ward discovered, in his study of Boston and New York in the 1890s, that some immigrants who lived in high density areas did not experience high mortality, and he cautions against

[36] The correlation between density and foreign-born in 1890 was +.53, whereas a weak negative relationship (−.20) existed between density and German-born. The negative correlation got stronger by 1900 (−.45, significant at $p<.05$), suggesting that Germans achieved even greater economic gains by the turn of the century.

equating immigrant status and congestion with high mortality.[37] Thus, although relationships existed between congestion and immigrant status and between congestion and death, these relationships only partially explain disease patterns and mortality. They do not tell the whole story.

The complexity of urban mortality is also evident in Milwaukee's experience with diphtheria. Death rates from diphtheria do not correlate with density in 1890, nor do they seem to have been influenced by the presence of foreign-born. Yet a fairly strong correlation connects German-born with diphtheria in 1890. This correlation grew stronger during the 1890s, so that by 1900 a statistically significant relationship existed between German ethnicity and diphtheria deaths. (See Table 1-5.) Since deaths from diphtheria apparently were not related to density or percentage of foreign-born, we must look to factors within the German community to explain its proportionately high disposition to death from the disease. Perhaps German resistance to

TABLE 1-5
Diphtheria Death Rates

	1890	1900
Density-within-dwelling	− .01	− .459
% Foreign-born	+ .26	+ .194
% German-born	+ .43	+ .605[a]

[a] $p < .01$.

[37] David Ward, "The Internal Spatial Structure of Immigrant Residential Districts in the late 19th Century," *Geographical Analysis* 1 (1969): 337-353. See also his *Cities & Immigrants: A Geography of Change in Nineteenth Century America* (New York: Oxford University Press, 1971). Ward noted that Jewish immigrants living side by side with Italians in congested wretched housing experienced low mortality while Italians suffered high rates, and he concludes that cultural adaptation explains part of the difference. Immunities gained from previous urban experience probably contributed to the difference as well. I would like to thank David Ward for his help in clarifying my thinking about Milwaukee mortality patterns. See also Robert M. Factor and Ingrid Waldron, "Contemporary Population Densities and Human Health," *Nature* 243 (1973): 381-384.

health-department regulations contributed to this situation. We know that many German parents refused to inoculate their children against smallpox; it is possible that some of the same fears greeted health-department efforts to control diphtheria and to administer antitoxin. The increasing gap between German and other immigrant groups with regard to deaths from diphtheria may have been created by different cultural reactions to the treatment. Adverse urban environments may have contributed to high mortality, but from this example it is clear that other factors also need to be considered for their impact on disease-specific mortality.

The limitations of congestion for explaining disease conformations are also evident in the case incidence of scarlet fever. A greater number of cases existed in the less congested parts of the city, indicating that crowding within dwellings did not exacerbate this disease. Density also did not correlate significantly with tuberculosis or pneumonia in 1890 and 1900. (See Table 1-6.)

Thus density and nativity predicted mortality from some nineteenth-century diseases and not from others. However, there are some compelling reasons for not discounting the general contribution of these two factors, despite their variability. First, other components that one would want to consider for various urban groups—such as immunity levels, availability of medical services, sanitation, and water, occupation, religion, breast-feeding data, and income lev-

TABLE 1-6

Density-Within-Dwelling

	1890	1900
Case rate-scarlet fever	n.a.	−.658[a]
Deaths from diphtheria	−.01	−.459
Deaths from tuberculosis	+.22	+.155
Deaths from pneumonia	+.38	+.363
TOTAL DEATHS	+.65[a]	+.514

[a] $p < .01$.

34

TABLE 1-7

Smallpox Reported Cases

	1868	1871	1873	1876	1894
American	n.a.	n.a.	41	n.a.	84
German	419	568	469	182	516
Polish	—	n.a.	21	232	386
Total	501	744	616	495	1074

n.a. = not available.

els—are not available or quantifiable on the ward level. Second, impressionistic information adds credence to the importance of ethnic identity and living conditions. Health officers often argued that German and Polish immigrants encountered the greatest risk of premature death in Milwaukee. Newspapers supported the theory that disease rampaged through immigrant neighborhoods "among a class whose antipathy to soap and water and other requisites of decency was proverbial."[38] Of course, public officials had strong reasons to point their fingers away from their own native-born constituency, but we need not overlook their observations in our efforts to temper them.

Health statistics do reveal an ethnic component to epidemic disease patterns in Milwaukee. Wards heavily populated with German and Polish immigrants and their families reported heavy disease casualties during the various smallpox epidemics. (See Table 1-7.) The fourteenth ward, on Milwaukee's far southwest side, frequently experienced the most devastation by disease of any ward in the city. Urban services took longer to reach this poor Polish ward than anywhere else; street paving, water pipes, and sewers lagged significantly behind settlement, whereas in other sections of the city urban services either preceded or accompanied new settlement. Houses in the fourteenth ward were the smallest in the city, but held the largest number

[38] *Sentinel*, November 20, 1876, p. 8. See also MHD, *Annual Report*, 1876, pp. 44-46.

TABLE 1-8
Persons-per-Dwelling in Milwaukee Wards

	1890	1900	1910		1890	1900	1910
1	6.32	5.86	6.09	13	5.71	6.10	5.91
2	6.82	6.46	7.18	14	7.30	8.73	8.96
3	6.43	7.32	10.07	15	6.25	5.60	5.12
4	6.29	6.93	7.85	16	5.43	5.13	6.01
5	5.80	5.74	6.20	17	5.72	6.05	6.06
6	6.22	6.16	5.83	18	7.24	6.83	6.07
7	6.14	7.17	7.86	19		6.25	5.57
8	5.66	5.32	5.32	20		5.70	5.35
9	6.21	6.06	6.30	21		5.83	5.57
10	6.04	5.81	5.87	22			5.65
11	6.55	6.58	6.43	23			5.23
12	6.55	6.48	6.45				

of persons of any ward. (See Table 1-8.) Roger Simon concluded in his analysis of 1905 housing in the fourteenth ward that, "although the transportation innovations of the late nineteenth century made it possible for immigrant unskilled laborers to live in new homes at the fringe of the city, and to aspire to home ownership, the urban environment which resulted was but a limited improvement over the congested conditions of the inner city wards."[39] The physical characteristics of this ward helped to define the quality of life and contributed to the excess of mortality experienced by the inhabitants.

This disease profile of Milwaukee indicates that crowding and nativity influenced mortality and that many other fac-

[39] Simon, p. 250. Although Simon concluded that the immigrants made the decision to postpone urban services, the situation was not entirely of their makng. To be near their places of employment, unskilled Polish laborers lived in south-side wards. Desirous of eventually having their own homes, many home owners rented rooms to other families to help to pay the mortgages, thus crowding their facilities beyond capacity. Since residents had to pay a portion of street improvements, they could not immediately afford to bring sewer and water pipes and paving into their communities. The city would not improve a section without the matching commitment from residents. Therefore, victims of poverty could not install some of the services that would have aided their plight.

36

tors, still unidentified or speculative, also contributed their
share. Immigrant status increased the risk of early death
from some diseases, while in others it may have provided
protection. Adverse urban conditions, such as crowding
and lack of sanitary amenities, helped to explain disease
patterns. Behavior of different ethnic groups, dietary and
sanitary habits, available medical care, advancing public
health measures, and a variety of other factors also defined
disease contours in urban America. Although we can only
partially explain why and how cities suffered because of
disease, it is clear that, for a combination of complex reasons, cities in the nineteenth century were relatively dangerous places in which to live. Rural areas in Wisconsin,
for example, for all their health problems, boasted a health
record superior to the urban areas until well into the twentieth century.[40] (See Figure 5.)

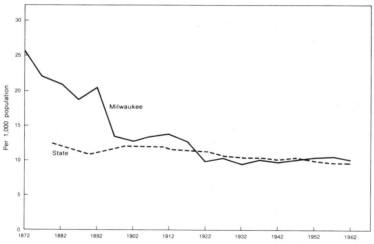

Figure 5. Wisconsin and Milwaukee Mortality Rates,
1872-1962. Courtesy of University of Wisconsin Press.

[40] Further exploration of the urban-rural differential in Wisconsin is
Judith Walzer Leavitt, "Health Care in Urban Wisconsin: From Bad to
Better," and Dale Treleven, "One Hundred Years of Health and Healing

37

Although German and Polish immigrant wards frequently suffered the highest mortality rates, disease incidence and devastation spread throughout Milwaukee. All city wards suffered from infectious diseases. The lack of concentration of cases of diphtheria, scarlet fever, typhoid fever, and tuberculosis indicates that no one was safe from urban diseases or free from the anxiety of premature death. Some were at greater risk than others, but death rates were not low in any ward in nineteenth-century Milwaukee.

Because cities posed grave threats to health and because health problems in turn menaced city growth, urban leaders repeatedly tried to turn back the tides of disaster. Rarely certain about exactly how to conquer this intolerable situation, Milwaukeeans falteringly tried various ways of improving their health statistics after 1867. Circumstances in the new immigrant wards remained bad when compared to the rest of the city, yet even in the most congested wards in the city, mortality decreased by the turn of the century. In the fourteenth ward, with the highest mortality, the death rate decreased from 25.1 to 15.7 per thousand population in the twenty years from 1890 to 1910. (See Table 1-9.) Child mortality decreased, and life expectation rose.

In part this general improvement was a function of expanding urban services. Although immigrant outlying areas were the last to receive such services as city water, sewer pipes, and street paving, these services gradually spread throughout the city at the end of the nineteenth century. (See Table 1-10.) The increased availability of water and waste disposal allowed people to assume daily habits that discouraged the rapid spread of disease. Changes in life style and diet, impossible to measure, undoubtedly contributed to improving statistics, as did the activities of the Board of Health. Initially created only in times of crisis and abandoned rapidly thereafter, health boards did not have

in Rural Wisconsin," in Ronald L. Numbers and Judith Walzer Leavitt, eds., *Wisconsin Medicine: Historical Perspectives* (Madison: University of Wisconsin Press, 1981).

TABLE 1-9

Death Rates in Milwaukee Wards

Ward	1870[a]	1879[b]	1890	1900	1910
1	18.0	16.1	15.3	10.94	11.30
2	23.9	13.6	15.7	11.88	11.73
3	27.6	17.7	19.1	12.51	12.00
4	21.3	14.2	14.9	15.43	12.66
5	20.8	18.2	15.8	12.54	11.64
6	24.2	13.8	15.9	11.38	11.56
7	14.6	11.7	13.1	11.57	8.43
8	27.6	19.5	20.2	10.33	10.80
9	28.4	17.5	17.5	13.26	10.81
10		17.3	16.3	11.01	10.10
11		26.3	20.2	12.09	11.66
12		21.3	19.6	14.18	11.37
13		13.9	18.9	13.58	10.65
14			25.1	16.14	15.73
15			14.1	9.04	10.93
16			13.3	10.66	14.41
17			15.4	12.28	11.93
18			15.5	14.13	12.58
19				8.94	13.33
20				16.02	11.73
21				12.28	12.61
22					11.82
23					12.33
City	23.1	21.0[c]	18.3	13.88	13.90

[a] 1 April 1870 to 1 April 1871.
[b] Mortality statistics by ward not available for 1880; rate based on 1880 population.
[c] City rate 1880 figures.

time or resources for long-term improvements. But after 1867 the permanent health officer, with a growing staff and budget, stimulated changes in city life that ultimately led to improved health standards. Public officials encouraged, obtained, and maintained control over food production and distribution in the city, lowering risks from contaminated milk, meat, and baked goods. City officials paid particular attention to sanitation, waging campaigns to improve garbage collection and disposal, water sources, sewage disposal, and privies. They monitored infectious dis-

TABLE 1-10

Ratio of Water Taps and Sewer Connections to Users

	1880	1890	1900	1910
Persons per water tap	16.78	12.04	6.87	5.85
Dwellings per water tap	2.73	1.93	1.1	1.26
Persons per sewer connection	n.a.	11.98	n.a.	6.69
Dwellings per sewer connection	n.a.	1.93	n.a.	1.44

n.a. = not available.
Source: Milwaukee, Department of Public Works *Reports* 1880, 1890, 1900, 1910 (Milwaukee, 1881, 1891, 1901, 1911).

eases and developed increasingly effective measures, including isolation, disinfection, hospitalization, and immunization, to control their spread. While precise measurements of the relationship between these measures and improving city mortality cannot be made, many of the health improvements of this period can be attributed to the lowered exposure to air- and water-borne diseases obtained through public health activities. The Milwaukee data substantiate at a local level Thomas McKeown's thesis that curative medicine was less important than hygiene, life style, and food quality in explaining decreasing mortality and increasing health. This study confirmed that during the late nineteenth century urban sanitation, food controls, disease prevention tactics, and cultural habits were major determinants of urbanites' health.[41]

The processes by which the city health department assumed its control are the focus of this study. The course of health reform was not a sure road toward progress and perfection. Health officials spent considerable energy on projects that had little effect on improving life and health standards. They argued among themselves and rarely worked with the full support of the medical and political communities. Only in times of acute crisis did the health

[41] McKeown, *The Role of Medicine*. See also Meeker, "The Improving Health of the United States."

department act with confidence and push its programs with vigor. While many factors prompted city residents, physicians, and politicians to support public health efforts, the obstacles to reform were legion. The economic motivation to maintain the status quo, although sometimes favorable to reform, repeatedly hampered efforts to change. Physicians frequently missed opportunities to advance their cause, because of internal disagreements or because they had difficulty manipulating the political machinery. Some health officers used their position in the city government to further private ambitions. Although "heroes" and "villains" emerge from this study of Milwaukee health reform, they are not always on opposite sides of the health department desk.

Through all the confusion associated with attempts to effect change in the rapidly expanding urban environment of the late nineteenth century, a coalition of health reformers emerged to support programs that directly or indirectly led to greater health among the Milwaukee citizens. By 1910 the city stood among the seven American cities showing the lowest death rates.[42] In 1930 Milwaukee won first place in the national Health Conservation Contest, an accolade that in part reflected the contributions of the health reformers.[43]

[42] *Sentinel*, December 7, 1911.

[43] John P. Koehler, "Acceptance Address," *American Journal of Public Health* 20 (1930): 635-636. Neither the contest judges nor Milwaukee officials used the term "healthiest" to describe the first-place victory. For the book title, I have interpolated that the repeated winner of the national health contest can be described as healthiest.

CHAPTER TWO

The City Health Department

Milwaukee always made some provision for its sick poor. In the years before the establishment of the permanent board of health in 1867, the city supplied physicians to the almshouse or to the local hospital to care for poor city residents. In times of epidemics it created temporary boards of health to deal with emergencies, allocating money and authority to a group of physicians selected by the Milwaukee City Medical Association. Although health-related expenditures were not a major part of the city budget, Milwaukee assumed ultimate responsibility for those activities beyond the purview of private charity.

Individual physicians and private hospitals provided some charity care. Milwaukee offered a wide range of healing services within the private sector. The majority of practitioners were regulars or allopaths, variously trained in the country's numerous medical schools or by apprenticeships. The quality of their training, and no doubt of their practices, ranged from excellent to extremely marginal, as did their economic success. In addition to its regularly trained physicians, Milwaukee boasted a large number of sectarian practitioners, who also represented diverse training experiences and quality. Most numerous among the sectarians were the homeopaths, followers of the German physician Samuel Hahnemann, who treated patients with infinitesimal drug doses. Fully one-third of Milwaukee's doctors were homeopaths during much of the late nineteenth century. Hydropathic and eclectic doctors also practiced in Milwaukee. The differences among available kinds of medical care, while important to individuals, only rarely intruded into public health concerns, because both regular and sectarian physicians participated in health department activities in the nineteenth century.[1]

The regular Milwaukee City Medical Association, formed in 1845, displayed a collective sense of public responsibility. Their first charter demanded members' vigilance "for the welfare of the community," urging them to "give counsel to the public . . . [in] public hygiene" and to "face the danger" willingly if pestilence struck.[2] Although these physicians individually recognized their duty to respond to health emergencies, as a group they responded only when the city paid for their services. The association agreed in 1848, for example, to send physicians to attend the "Almshouse, Pesthouse and wards of the city in every pauper care, and also furnish a board of health" for "the sum of six hundred dollars to be paid quarterly."[3] In 1852 the fee rose to $700 per year.[4]

---

[1] Elizabeth Barnaby Keeney, Susan Eyrich Lederer, and Edmond P. Minihan, "Sectarians and Scientists: Alternatives to Orthodox Medicine," in Ronald L. Numbers and Judith Walzer Leavitt, eds., *Wisconsin Medicine: Historical Perspectives* (Madison: University of Wisconsin Press, 1981), pp. 47-74. See also Ronald L. Numbers, "Do-It-Yourself The Sectarian Way," in *Medicine Without Doctors: Home Health Care in American History*, ed. Guenter B. Risse, Ronald L. Numbers, and Judith Walzer Leavitt (New York: Science History Publications, 1977), pp. 49-72, and Martin Kaufman, *Homeopathy in America: The Rise and Fall of a Medical Heresy* (Baltimore: Johns Hopkins Press, 1971).

[2] *Constitution, By-Laws and Code of Ethics of the Milwaukee City Medical Association*, 1847, p. 22.

[3] MCMA Minutes, June 22, 1848, p. 33. See also June 15, July 6, 1848. Others have claimed that the MCMA provided a free board of health. See George A. Dundon, comp. "Health Chronology of Milwaukee," unpublished typescript in the Milwaukee Health Department Library, 1848 entry; and G. Kasten Tallmadge, "Confutations I," *Milwaukee Medical Times* (November 1939), p. 49. The MCMA allocated jobs among its members either by lottery or through bidding. MCMA minutes, July 11,

The early boards of health created under these arrangements dealt only with epidemic emergencies. The city appointed its first boards during the smallpox epidemics of 1843 and 1846 and others during the cholera scares of 1848 to 1854.[5] The boards located and isolated the sick, vaccinated (in smallpox epidemics) to protect the well, and provided treatment in isolation hospitals for those who could not be taken care of at home. From 1855 to 1866 the mayor and the common council constituted the board of health, but they rarely, if ever, met in that capacity.[6]

Although Milwaukee survived these sporadic attempts to control the ravages of infectious diseases, many physicians and business people realized that failure to adopt a more comprehensive policy hampered city growth. They realized, too, through lessons in sanitation and prevention learned on the Civil War battlefields, that diseases spread rapidly in crowded environments but that some of the resulting devastation could be prevented. The dominant miasmatic theory of disease, which relied on the observable connection between filthy conditions and disease prevalence, posited that rotting organic material and dirt adversely affected the atmosphere and permitted the bad air to carry sickness. Cleaning up the dirty environments would, according to this theory, help to alleviate the menace of disease. This filth theory of disease, present in American

1848. If by bidding, any money left over went into the association treasury. Records of the Medico Chirurgical Club, January 28, 1851, February 11, 1851. This association had a short life during a lapse of the MCMA.

4 Records of the Medico Chirurgical Club, December 29, 1851.

5 Louis Frank, *The Medical History of Milwaukee 1834-1914* (Milwaukee: Germania Press, 1915), p. 184; MHD, *Annual Report*, 1871, pp. 9-10; MCMA Minutes February 17, 1848, p. 17; March 2, 1848, p. 18. See also Walter Kempster and Solon Marks, "General Sanitation and the Health Department," in Howard Louis Conrad, ed., *History of Milwaukee from Its First Settlement to the Year 1895* (Chicago, n.d.), I, 250-254.

6 Frank, *Medical History*, p. 184; MHD, *Annual Report*, 1871, pp. 10-11; MCMA Minutes, December 5, 1855; April 6, 20; May 4, 1859.

medical thought from the eighteenth century, grew in popularity during the middle of the nineteenth century.[7] Realizing that some diseases could be prevented by sanitation efforts and fearing that Milwaukee's future depended on avoiding periodic epidemic destruction, three hundred "prominent" citizens petitioned the legislature in 1864 for a change in the city charter, to remove the city agencies responsible for health, fire, and police protection from local control and allow them to be appointed at the state level. The state legislature, responding to political pressure to keep power within the city, defeated the measure.[8]

Because of this citizen attempt to wrest control from local officials, the aldermen actually met as a board of health in 1865 and 1866 to provide "for the more effectual protection of the health of the city."[9] No action resulted from these meetings, however, and continued public dissatisfaction with a board that acted only in response to emergencies led in 1867 to a second attempt to create a permanent board of health in the state legislature.[10] The example of the newly created Metropolitan Board of Health in New York City helped to convince Wisconsinites to provide a similar body for their state's largest city. In April 1867 the legis-

[7] See, for example, John Duffy, *Public Health in ·New York City 1866-1966* (New York: Russell Sage, 1974) and Philip Jordan, *The People's Health: A History of Public Health in Minnesota to 1948* (St. Paul: Minnesota Historical Society, 1953). For more on the miasmatic theory of disease, see Richard Harrison Shryock, *The Development of Modern Medicine*, reprint edition (Madison: University of Wisconsin Press, 1979) and George Rosen, *A History of Public Health* (New York: MD Publications, 1958).

[8] *Sentinel*, March 23, 1864. I was unable to uncover a copy of the petition, and so the 300 individuals remained unidentified. The *Sentinel* maintained that they acted independently of party considerations. See also, *Sentinel*, April 1, 2, 1864.

[9] *Sentinel*, August 1, 1865, p. 1; April 26, 1866, p. 1; October 2, 1866, p. 1.

[10] Laurence M. Larson, *A Financial and Administrative History of Milwaukee* (Madison, 1908), pp. 36-37.

lature approved a compromise measure that allowed the mayor of Milwaukee, instead of the governor, to appoint a permanent five-member board, "whose duty it shall be to examine into and consider all measures necessary to the preservation of the public health in the city of Milwaukee."[11]

In the spirit of compromise, the legislature put no limitation on the composition of the board, allowing common council members to serve if the mayor wished to appoint them. Conceivably, the mayor could have continued in the old pattern, except that the new board had broader powers than the old. The state empowered the permanent board to remove unhealthful nuisances, cleanse and purify public areas, destroy "bedding, clothing, putrid or unsound beef, pork, fish, hides or skins" deemed dangerous, penalize any who refused to obey orders, occupy buildings for use as hospitals in times of emergency, and require physicians to report contagious diseases. Instead of reacting only when epidemics threatened, the new board could regularly set policy to prevent disaster. Even though the Wisconsin board was not as independent of local politics as its New York cousin, the potential for meeting Milwaukee's health and prevention needs now existed.[12]

In the years following its creation, the health board underwent various structural modifications. At first the five-member board elected the medical health officer from among its own group. In 1874 the board reverted to a committee of aldermen, and in 1878 the health officer became health

[11] *Private and Local Laws Passed by the Legislature of Wisconsin in the Year 1867* (Madison: Atwood and Rublee, State Printers, 1867), Chapter 595, pp. 1283-1287; MHD, *Annual Report*, 1871, pp. 11-12; *Sentinel*, May 6, 1867, p. 1.

[12] For the New York example see John Duffy, *A History of Public Health in New York City 1607-1866* (New York: Russell Sage, 1968), and Gert H. Brieger, "Sanitary Reform in New York City: Stephan Smith and the Passage of the Metropolitan Health Bill," *Bulletin of the History of Medicine* 40 (1966): 407-429.

commissioner, heading his own department and working through an aldermanic health committee. His term of office increased from two to four years in 1882. After 1895 city civil-service rules governed most department employees.[13] None of these changes significantly altered the basic power structure: health officials were beholden to the mayor for appointment and subservient to the common council for budgetary allocations. All health policy initiated by the medical people had to be approved by the politicians. Because of the dependent relationship, health officials had to learn to work with political restraints, to compete with other city departments for funds and attention, and to manipulate the city bureaucracy for their own benefit. Health officers varied in their abilities or desires to meet the political requirements of their jobs, and health policy innovations frequently succeeded or failed, depending on the doctors' political sophistication.

Dr. James Johnson, the first health officer, held the office for ten years and was one of the most accomplished health politicians ever to serve Milwaukee. (See Table 2-1.) He had practiced medicine in Milwaukee since 1844, was a charter member of the Milwaukee City Medical Association, and participated actively in city affairs. Johnson, whose gregarious personality enlivened his liberal politics, advocated free schools and served as an independent alderman before becoming the city's sanitation officer, and he carried his political experience and boundless energy to his new tasks.[14] He worked virtually alone, meeting only occasion-

---

[13] Dundon, "Health Chronology," entries under appropriate years. On civil-service reform see Thelen, *New Citizenship*, pp. 161-164.

[14] MHD, *Annual Report*, 1871, pp. 11-12; *Sentinel*, May 6, 1867, p. 1. For biographical material on James Johnson, see Frank, *Medical History*, p. 12, F. M. Sperry, *A Group of Distinguished Physicians and Surgeons of Milwaukee* (Chicago: J. H. Beers and Co., 1904), or United States Works Progress Administration, "Field Notes for a Biography of Dr. James Johnson" in the Archives of the Wisconsin State Historical Society (WSHS). Dr. Enoch Chase, another pioneer Milwaukee physician, noted Johnson's

TABLE 2-1
Milwaukee Health Officers

Health Officer	Years in Office	Medical School	Year of Graduation	Type
James Johnson	1867-1877	Berkshire Medical College	?	Regular
Isaac H. Stearns	1877-1878	National Medical College	1860	Regular
Orlando W. Wight	1878-1881	Long Island College Hospital	1865	Regular
Robert Martin	1881-1889	Starling Medical College	1875	Homeopathic
U.O.B. Wingate	1890-1894	Dartmouth Medical School	1874	Regular
Walter Kempster	1894-1898	Long Island College Hospital	1864	Regular
H. E. Bradley (acting)	1895	University of Buffalo	1887	Regular
F. M. Schulz	1898-1906	Rush Medical College	1896	Regular
G. A. Bading	1906-1910	Rush Medical College	1896	Regular
W. C. Rucker	1910	Rush Medical College	1897	Regular
F. A. Kraft	1910-1914	American Medical College-St. Louis	1894	Eclectic
George C. Ruhland	1914-1924	Wisconsin College of Physicians & Surgeons	1904	Regular
John P. Koehler	1924-1940	Marquette University Medical Department	1911	Regular
Edward Krumbiegel	1940-1973	Marquette University School of Medicine	1935	Regular
Constantine Panagis	1973-	Marquette University School of Medicine	1946	Regular

ally with the board, whose membership varied, and carved out interest areas that became models for his successors. For $1,000 a year, Johnson coped with the daily health duties and maintained the "pleasantest" relations with his board of aldermen.[15] He travelled to Europe in 1874 at his

---

booster spirit in a speech he made before the Old Settler's Club on July 4, 1872: "Under the benignant and guiding influence which you created, you have seen Milwaukee become the first primary wheat market in the world, the fourth pork packing city in the Union, the second commercial city on Lake Michigan, the seventeenth in population, and according to Dr. Johnson, the healthiest American city." Cited in WPA Biog. notes from James S. Buck, *Pioneer History of Milwaukee*, vol. 1, p. 253.

[15] *Proceedings of the Common Council*, May 4, 1874, pp. 14-15. MHD, *Annual Report*, 1875, pp. 9-10.

own expense to investigate new directions for sanitary activity. Many who had reservations about his policies admired Johnson's commitment to solving Milwaukee's critical health problems. But for all the positive reactions he generated within city hall and outside of government, Johnson still found it difficult to increase vaccinations, sanitary controls, or even his own salary, and his career as health officer illustrates the frustrations that became endemic to the job. Health reforms cost money, and politicians remained reluctant to allocate the necessary funds until situations proved so desperate that they had no choice. Not until it became politically expedient after the turn of the twentieth century regularly to support disease prevention did aldermen show positive reactions to the health officers' programs.

The careers of other physician health officers also reflected this pattern of political restraint. Dr. Orlando W. Wight, Milwaukee's first health commissioner, served from 1878 to 1881, when he left the city for a similar job in Detroit. A former lawyer and Unitarian minister, Wight addressed health problems with a crusading spirit. His haranguing speeches and press interviews lost him some support among the public, but his strong beliefs added to the fervor and devotion with which he attacked the city's health problems. He was the first public physician to uncover the unsanitary conditions of Milwaukee's dairies, and he led an unsuccessful battle to improve the quality of milk in the city. Wight's reputation as a "windbag" and his lengthy written and oral reports to the common council sometimes hampered his effectiveness in creating new controls over health nuisances.[16] But Wight made good newspaper copy and his personal visits to 227 foul cow stables dramatically advertised the city's sanitation hazards. His decision to leave Milwaukee was probably influenced by his frustrations with the common council, although he was not entirely ineffec-

[16] *Daily News*, June 10, 1879. For biography and photograph of Dr. Wight, see Frank, *Medical History*, pp. 50-51.

tive. During Wight's administration the health office added two medical assistant commissioners, two sanitary inspectors, and an office secretary.[17]

Dr. Robert Martin, who served as Milwaukee's health commissioner from 1881 to 1889, was the only homeopath to hold the job, despite the large number practicing in the city. Originally appointed only to fill out Wight's term, the Scotsman hung on to the job for two full terms of his own, overriding objections from many quarters that he lacked the "necessary qualifications."[18] Martin tried to continue many of Wight's programs (he had been Wight's assistant), but he encountered even greater resistance than had Wight from council members and the public. He did not fight back as Wight had done. Although active in trying to create solutions to Milwaukee's garbage and milk problems, Martin did not devote his full energies to the job. He never issued a single health department annual report, and he was almost impeached for using city funds inappropriately.[19] His tenure in office came to an end when Democrats swept the city in 1890 and appointed Dr. U.O.B. Wingate to office.

A pillar of the regular medical profession, Dr. Wingate served four years as health commissioner and brought order and propriety to the health department. His "sterling qualities as a highly cultured gentleman" served him well with the doctors, and he won the close cooperation of the profession in his public endeavors.[20] Wingate also held the

[17] MHD, *Annual Report*, 1879.

[18] *Sentinel*, October 1, 1881. The German aldermen who made the accusation did not say that those qualifications were that he be German or a regular, but they named no concrete objections to Martin. *Sentinel*, October 3, 12, 18, 1881. The regular medical community did not support Martin's appointment. *Sentinel*, October 12, 1881.

[19] *Sentinel*, July 21, 1888. See also August 22, 1888. For biographical data on Martin, see Frank, *Medical History*, p. 56. Interestingly, New York City issued no annual reports in the 1880s either. See John Duffy, *A History of Public Health in New York City 1866-1966* (New York: Russell Sage, 1974), pp. 66-67.

[20] Frank, *Medical History*, p. 74.

respect of the aldermen and succeeded in getting them to pass legislation on milk and privy vaults. Helped along by a threat of cholera, he increased the powers of the health commissioner to act unilaterally in times of epidemic crisis. His was an active administration, marked by efforts to bring professional standards and ambiance to the health department. When Wingate left to become Secretary of the Wisconsin State Board of Health in 1894, this New England-bred commissioner turned over to his successor a well-functioning and expanded department: three medical assistants, a registrar of vital statistics, eighteen sanitary inspectors, a clerk, and a stenographer.

Milwaukee's stormiest commissioner turned Wingate's order into political confusion. Dr. Walter Kempster, a London-born, internationally recognized psychiatrist, received the reform Republican mayor's appointment in 1894 and promised to lift the department out of political influence through civil-service reform. However, when he refused to appoint loyal members of his own party to office during the severe economic depression, Kempster found himself in a political quagmire even deeper than any of his predecessors. He tried to stay above politics, but felt himself pulled into the fray by his refusal to work within the traditional system. At the height of a smallpox epidemic, when citizens rebelled against Kempster's attempts to control the disease, the common council impeached him and threw him out of office.[21] This event and all the publicity it engendered tarnished the reputation that the health department had achieved under Wingate. Kempster managed to get his job back in 1896 through a court battle, and before finally leaving office in 1898 he installed a laboratory to analyze systematically the quality of city milk and water and expanded the size of the department staff.[22] Certainly one of Milwaukee's most sensational public figures, Walter

---

[21] For details on this action, see Chapter Three.

[22] The laboratory was approved in October 1893, but it did not begin systematic work until 1896. MHD, *Annual Report*, 1893, p. 21.

Kempster fell victim to the political pitfalls of the job of health commissioner.

Dr. F. M. Schulz took over from Kempster and served two terms, from 1898 to 1906. Mayor David S. Rose appointed Schulz above aldermanic protests from Rose's own party that Schulz was "incompetent."[23] Ignoring civil-service ice rules about employees, Schulz gradually dropped Republicans from the health department and appointed fellow Democrats in their stead.[24] Known for his "extravagance" in office, he expanded the staff and laboratory functions of the department.[25] There were no crises during his administration, and Schulz never became the controversial doctor that Kempster had been. He continued department work, expanded activity in many areas, and lauded his own achievements in his annual reports. However, the stigma of corruption and inefficiency that finally defeated Mayor Rose also followed Schulz out of office. His successor, in a politically motivated statement, decried Schulz's "lack of proper discipline," his "lack of business methods," and his "extravagance of various kinds."[26]

Dr. Gerhard A. Bading, who served from 1906 to 1910, espoused the business-efficiency ethic of the reform Republican mayor who appointed him. Bading made sweeping cuts in expenditures, including some for department personnel who he claimed served only because of "political influence." In his second year on the job Bading handed the city treasurer a cash balance of over $18,000.[27] At the same time he expanded the activities of the health department to include licensing food establishments, inspecting school children, and providing tuberculin testing for safer

[23] *Sentinel*, November 22, 1898.
[24] *Sentinel*, August 2, 1898.
[25] See, for example, *Sentinel*, February 2, 1907, and the MHD, *Annual Report*, 1906, p. 11.
[26] MHD, *Annual Report*, 1906, p. 11.
[27] MHD, *Annual Report*, 1907, p. 11.

milk. The first commissioner born in Milwaukee, Bading ran an efficient and tight health department.

Supported by a responsive Socialist common council, Dr. F. A. Kraft continued to expand the scope of governmental controls over health during his administration from 1910 to 1914. After a demonstration child welfare station in the fourteenth ward illustrated the effects that intervention could have on health by cutting infant deaths in half in one year, Kraft incorporated many such community projects within the health department. He expanded public education programs and helped to set the directions maintained by the three commissioners who served until the 1970s: George C. Ruhland (1914-1924), John P. Koehler (1924-1940), and Edward R. Krumbiegel (1940-1973). Long tenures in office since 1914 have made it possible for health officials to build comprehensive health programs and have taken the office away from its earlier association with politics.

The job of health commissioner was never relaxed or lucrative, yet it attracted physicians who worked hard to promote the city's health. The web of politics engulfed the office until well into the twentieth century, and some commissioners could not escape its influence. Although they often did not meet the demands of the job, they failed more from lack of ability than from malice, more from political considerations than from medical ones. They were a varied group and espoused different purposes, politics, and medical beliefs; but they shared frustrations born out of trying to effect change within the municipal structure. Their common experiences led to a continuity of health policy in the department as each official tried to build on the accomplishments of the physicians who preceded him, even if publicly they spoke against previous administrations. The job itself—filled with daily frustrations and unmet expectations—welded their differences into one continuous struggle.

And it was a struggle from the beginning in 1867, when

Dr. Johnson promised to take "energetic measures" to preserve the city's health.[28] Except during epidemic crises, when the health officer assumed extra authority, Johnson and his successors relied heavily on their power to abate nuisances in their efforts to make Milwaukee a healthier place. Nuisance complaints—about overflowing privies, uncollected slops, offensive smells, and extraordinary filth—poured into the health office every day. Health officials investigated each offense, served notice to those responsible, and revisited to make sure the situation had been remedied.[29] Through nuisance abatement, health departments were able to expand their authority over conditions the doctors thought most responsible for high urban sickness and death.

One of the earliest nuisances to catch Johnson's attention was the odor emanating from the slaughter houses and meat-processing plants around the city. In 1869 and 1870 the health officer visited local slaughter houses and ordered them to keep their establishments "as free from offensive odors as the circumstances and nature of their business permits."[30] Johnson did not want to push them any harder because Milwaukee could not afford to discourage business people "who have added, and are still adding, so much to the wealth and prosperity of our city."[31] Johnson hoped residents could withstand the "slight inconvenience" of the "peculiar, and to some persons, offensive smell" in the interests of city prosperity.[32] However conciliatory these statements sounded, Johnson monitored the meat processors closely and also tried to regulate the conditions under which meat was marketed.[33] The common council consis-

[28] *Sentinel*, June 13, 1868, p. 1. See also August 24, 1867, p. 1; June 10, 1868, p. 1.

[29] MHD, *Annual Report*, 1870, p. 1.

[30] *Ibid.*, p. 6; *Sentinel*, June 16, 1869, p. 1.

[31] MHD, *Annual Report*, 1870, p. 6.

[32] MHD, *Annual Report*, 1872, pp. 36-38.

[33] *Ibid.*, pp. 11-16. MHD, *Annual Report*, 1873, p. 18; *Proceedings of the Common Council*, June 8, 1874, p. 52.

tently thwarted his hopes of gradually increasing controls by refusing to provide meat inspectors or to limit the parts of the city in which meat could be processed.[34]

Although slaughtering establishments dotted the city, the largest and most offensive were concentrated in the Menomonee valley directly west of the business district, where they dumped their wastes into Burnham's Canal, to be carried to the river. Distilleries also deposited their waste into the canal. In 1874 an inspection revealed that "the thick, inky, putrid water . . . [was] in a state of violent commotion, produced by the fermentation existing at the bottom. . . . The water, grains, cow manure, and other filthy matter was thrown by the power and explosive force of the gas generated many feet into the air . . . [resembling] some great subterranean explosive power."[35] Johnson worked to eliminate this public nuisance in the 1870s, but he came to realize before he left office that the only cure to the problem would be legislation outlawing slaughtering within the city.[36] Because of the economic benefit provided by the slaughter houses in the growing city he was unwilling to pursue such a course.

Dr. Wight was not nearly so cautious. He immediately tried to get the council to pass prohibitory legislation. When he failed in this attempt, Wight used his nuisance powers to restrict the disposal of offal and claimed sanitary improvements for his surveillance.[37] Dr. Martin went further and in 1888 convinced aldermen to let him restrict slaughtering to prescribed parts of the city. The butchers organized to fight the order, vowing to "continue killing as before," but the force of a sample arrest and fine brought the "marketmen" to compliance.[38] Dr. Wingate built on the work of his predecessor and further proscribed conditions

[34] MHD, *Annual Report*, 1874, p. 49.
[35] MHD, *Annual Report*, 1874, p. 34.
[36] *Ibid.*, p. 39. See also *Proceedings of the Common Council*, June 15, 1874, p. 60.
[37] MHD, *Annual Report*, 1879, pp. 250-252.
[38] *Sentinel*, March 30, July 21, 1888.

under which butchering could be practiced: after 1892 all establishments had to have water-tight asphalt floors and sewer connections.[39]

Enforcement of all of these regulations remained difficult and sometimes impossible in a health department short of money and personnel. So it was not surprising, when *The Jungle* raised the national consciousness about foul meat in 1906, that an investigation of local butchers in Milwaukee revealed conditions similar to those in Chicago.[40] Inspections of slaughtering and packing establishments uncovered manure piles in close proximity to slaughtering tables, dead rats mingled with fresh meat, and excessive filth.[41]

Public horror at these unsanitary conditions brought pressure for effective legislation. After 1907 the health department licensed all meat markets, and the council cooperated by doubling the department's inspection staff.[42] When the Socialists commanded the council in 1911, the aldermen went even further and passed an ordinance outlawing rendering from the city entirely—an action that James Johnson would have applauded.[43] But the council had second thoughts about the economic wisdom of pushing lucrative business out of the city and postponed enforcement. Under the threat of banishment, many rendering plants voluntarily installed new equipment to eliminate odors, an improvement both the health department and the common council were inclined to accept. Officials had cast off their initial reluctance to restrain business, but they still acknowl-

[39] *Common Council Proceedings*, November 13, 23, 1892.

[40] Upton Sinclair, *The Jungle* (New York, 1906).

[41] *Sentinel*, April 6, 1907. For further details on the council investigation, see: *Sentinel*, July 14, 1906, March 2, 28, April 3, 24, 1907; *Evening Wisconsin*, February 27, March 2, 4, 25, April 3, May 1, 11, 1907; *Journal*, March 2, 9, 1907; *Daily News*, March 4, 16, 27, April 8, 24, May 1, 1907; *Free Press*, March 27, April 26, May 1, 1907. See also the clippings box in the Local History Room of the Milwaukee Public Library.

[42] MHD, *Annual Report*, 1907, p. 19; MHD, *Annual Report*, 1908, p. 21.

[43] MHD, *Annual Report*, 1911, pp. 20-22; MHD, *Annual Report*, 1912, p. 44.

edged that business interests could not be ignored in the quest for a healthy sanitary environment.

A task more frustrating than regulating meat processing was trying to influence water and sewerage policy. Because both fell technically to the department of public works, health officers could only complain or suggest; they could effect no policy on their own. To influence water or sewage policy, the health commissioner had to lobby with the engineers as well as confront the aldermen.

Milwaukee provided its citizens with water only after the needs of nearly 100,000 people forced the issue. In 1868 the common council hired Chicago's engineer, E. S. Chesbrough, to plan a water system for the city. Three years later the aldermen hired engineer Moses Lane to implement Chesbrough's plans.[44] In 1874 Milwaukee opened its first municipal hydrants, and clean Lake Michigan water surged through the underground pipes to replace the familiar cisterns, wells, and water carts. To minimize the public debt thus incurred, the city charged homeowners a minimum of $4.00 a year to use the water. Homeowners had the additional expense of hiring plumbers to install the equipment and connect it to the city system. Many could not afford such luxury and continued to draw from the increasingly contaminated local wells. Health officer Johnson tried to convince the engineers that "it was a blind policy to put the water rates, on the first starting out, at so high

[44] E. S. Chesbrough, *Report on Milwaukee Water Works, submitted by E. S. Chesbrough, civil engineer, to his Honor the Mayor and the Committee on Water Works of the City of Milwaukee, October 28, 1868* (Milwaukee: Daily News Steam Book and Job Print, 1868). Copy in the WSHS. For material on Chesbrough, see Louis P. Cain, "Raising and Watering a City: Ellis Sylvester Chesbrough and Chicago's First Sanitation System," *Technology and Culture: The International Quarterly of the Society for the History of Technology* 13 (1972): 353-372; Larson, *Financial and Administrative History*, pp. 108-113; Charles Eliot Beveridge, "History of Water Supply in the Milwaukee Area," unpublished M.A. thesis, University of Wisconsin-Madison, 1959, pp. 1-3; Elmer W. Becker, *A Century of Milwaukee Water* (Milwaukee: privately printed, 1977); and MHD, *Annual Report*, 1871, pp. 38-39.

a figure," arguing that the poor, who benefited most from adequate and clean sources of water, would be the last to receive it.[45] He and his successors continued with no success to badger the city officials into lowering the rates for the poor.[46] Despite his reservations about distribution policy, Johnson shared the city's jubilation when the water works opened and boasted that, "There is no city in the Union better supplied with pure water than Milwaukee."[47]

To allow waste water to flow out of the city, Milwaukee next had to expand its sewerage system. Old and new sewers emptied untreated wastes into the city's three rivers, creating what Johnson called a "huge cesspool" in the middle of the city.[48] The "river nuisance" became a major public issue during the summer of 1879, when the rivers sent up an unprecedented "stench in the nostrils of the people."[49] In one of Health Commissioner Wight's longest epistles to the common council, which won him the title "great American stink investigator" from the press, he com-

[45] MHD, Annual Report, 1874, pp. 68-69; Larson, p. 111. By 1884, ten years after the municipal service opened, 110 miles of water pipes were buried under Milwaukee streets, but fewer than 10,000 customers took advantage of the benefits of lake water. For more statistics of water mains, see the Annual Reports of the Board of Public Works; Samuel Weidman and Alfred Schultz, The Underground and Surface Water Supplies of Wisconsin, Wisconsin Geological and Natural History Survey, 35, Economic Series No. 17 (Madison, 1915): 454-457; The Manual of the American Water Works (Engineering News, 1889-1890), pp. 469-470; 1890-1891, pp. 251-252; and Becker, A Century of Milwaukee Water, pp. 165-167.

[46] MHD, Annual Report, 1877, p. 19.

[47] MHD, Annual Report, 1876, p. 68. See also MHD, Annual Report, 1875, p. 44.

[48] MHD, Annual Report, 1873, pp. 31, 35-37. See also MHD, Annual Report, 1871, p. 38; Sentinel, August 18, 1873, p. 8; and Larson, Financial and Administrative History, p. 115.

[49] Milwaukee State Journal and South Side Advocate, July 10, 1879. See also Sentinel, August 11, 12, 13, 1879 and July 8, 1879; State Journal and South Side Advocate, August 2, 16, 1879; Daily News, August 9, 10, 12, 13, 16, 21, 22, 1879; and the Cream City Courier, August 16, 1879. The distillers claimed they sold their swill to milkmen or gave it away, denying their contribution to the river nuisance.

plained bitterly about the unhealthy nuisance, and newspapers and citizens joined him in protest.[50] The council responded feebly to the public outcry by hiring consulting engineers to examine Milwaukee's plight. But when the country's leading experts, E. S. Chesbrough, Moses Lane, and George Waring, recommended a system of intercepting sewers to relieve the rivers' pollution, the council refused to build even the minimal flushing tunnel. When the stench subsided with the passing of hot summer weather, the aldermen even resisted paying the fees of the consulting engineers.[51]

For seven more years the rivers continued their "filthy flow unvexed to the lake."[52] After another odor-ridden year in 1887, when the press daily vented the public impatience, the common council again rejected plans for expensive intercepting sewers to take wastes directly to the lake, instead adopting a compromise solution to build a flushing tunnel on the worst of the offending rivers.[53] By the time the American Public Health Association held its annual meeting in Milwaukee in November 1888, the Milwaukee River flushing tunnel had improved the quality and smell of river water, and the city proudly exhibited its benefits before the visitors.[54]

[50] *Daily News*, August 30, 1879. See also *Daily News*, August 23, 26, 1879.

[51] The engineers' report is in the MHD, *Annual Report*, 1879, pp. 243-249. The three men were the most famous engineers in America at the time. For information on George Waring, see James H. Cassedy, "The Flamboyant Colonel Waring: An Anticontagionist holds the American Stage in the Age of Pasteur and Koch," *Bulletin of the History of Medicine* 36 (1962): 163-176; and Richard Skolnik, "George Edwin Waring, Jr., A Model for Reformers," *New York Historical Society Quarterly* 52 (1968): 354-378. See also *Daily News*, September 30, October 2, 1879; February 3, 1880; and the *Sentinel*, January 31, 1880.

[52] *Sentinel*, August 7, 1887. See also letter to the editor, *Daily Review*, July 19, 1887; and the *Sentinel*, August 11, 12, 14, 16, 1887.

[53] *Sentinel*, August 26, 1887; August 31, October 23, 1887.

[54] George H. Benzenberg, "The Flushing Tunnel at Milwaukee," *Public Health: Papers and Reports* 14 (1889): 188-189. By 1888 Milwaukee owned 165 miles of sewer pipes.

Health officers began to wonder how increasing amounts of sewage in the lake affected the quality of the city drinking water. "How can the water of the lake be pure when so many filthy rivers flow into it?" asked Dr. Wight.[55] Leaked to the press, his worry became sensation. The *Daily News* found "death in our drink," and the more moderate *Evening Wisconsin* concluded that "a most unwholesome, unhealthy fluid" came out of the city water taps.[56] The quality of the city water continued to concern health officials and the public through the next three decades, periodically exploding as lively copy in the daily newspapers.[57] A new intake tunnel constructed in 1891 temporarily relieved the lake-water contamination, but both physicians and engineers agreed that the water would have to be filtered or otherwise purified and the sewage treated before they could relax their vigilance.

In 1896 the new health department laboratory began daily analysis of the city water. Scientists found well water "contaminated to such an extent that it is unfit for drinking."[58] Out of 118 wells investigated, officials designated the water from only 16 safe for ingestion. The contamination, inevitable in crowded environments, came from privy vaults and cow stables. Laboratory workers also found fault with the lake water, fearing that the 54 million gallons of raw sewage dumped into the lake each day led to increased incidence of typhoid fever in the city.[59] In 1905, for example, Health Commissioner Schulz, alarmed over rising numbers of intestinal infections, urged improved techniques of sewage disposal and water filtration, but to no avail.

The situation grew critical when the Kinnickinnic flush-

[55] MHD, *Annual Report*, 1878, p. 113. See also pp. 116, 121.

[56] *Daily News*, August 28, 1879; *Evening Wisconsin*, July 29, 1879.

[57] See, for example, *Sentinel*, May 7, 1881, p. 3; June 3, 19, 1887; September 25, 1887.

[58] MHD, *Annual Report*, 1896, p. 29; MHD, *Annual Report*, 1899, p. 11.

[59] MHD, *Annual Report*, 1897, 1898, 1899.

ing tunnel opened in 1907 and emptied more sewage into the lake every day. Increasingly successful at shutting down private wells, which left citizens no recourse but to use city water, the health department could not convince the engineers and council that the potential "harvest of calamity" warranted larger expenditures for water filtration and sewage treatment. Health Commissioner Kraft vented the frustrations of his predecessors: "The pouring of millions and millions of gallons of untreated sewage daily into the water upon which the people of this city depend for drinking and domestic purposes seems a practice so primitive, so unspeakably filthy and crude, that one is at a loss to explain [it]."[60]

After a typhoid epidemic in 1910, city council members finally recognized the importance of action and began sporadically treating the water with hypochlorite of lime, a practice made permanent after 1912. A breakdown in the chlorinating apparatus in 1916 led to a widescale bout of diarrhea in the city, which in turn precipitated a successful referendum on a bond issue for a sewage treatment plant. Milwaukee's "typhoid highball" became a thing of the past,[61] and the city's public physicians relaxed their surveillance over water and sewage policy.

Throughout the period when Milwaukee public doctors worried about drinking water, they also monitored the sanitary quality of the city's ice, an important consumer commodity in the years before electric refrigerators. Ice companies traditionally cut their ice from the three rivers that ran through Milwaukee. Increased sewage in the rivers after 1874 contaminated this source, so health officers set geographic limits to ice cutting, restricting it to regions above the entrance of sewage pipes. However, unless they actually caught people cutting ice from the parts of the river below the sewer entrances, health officers found it

[60] MHD, *Annual Report*, 1911, p. 94.
[61] Beveridge, "History of Water Supply," p. 17.

Figure 6. Ice harvesting in Milwaukee. Courtesy of
Milwaukee Public Museum.

hard to prove negligence. At least one company deliberately tried to deceive health officers by storing country ice in front of river ice so that when inspectors visited they saw only the clean supply.[62]

Johnson and Wight, who carefully watched the ice dealers, thought their inspections kept the dangerous river commodity off the market. But Dr. Kempster's laboratory analyses, begun in 1896, revealed Kinnickinnic sewage in some city ice. Decrying the "cupidity of man" that allowed people to sell "adulterated and unwholesome materials" to unsuspecting residents, Kempster cracked down hard on

[62] MHD, *Annual Report*, 1876, p. 86; *Sentinel*, February 24, 1877, p. 3; MHD, *Annual Report*, 1877, pp. 45-46. Ice for use in refrigeration only, such as for beer kept in kegs or barrels, was allowed to be cut below the sewers.

those selling "frozen sewage."[63] The incident revealed to Kempster the importance of the "utmost vigilance" on the part of public physicians in the daily monitoring of ice and other consumer commodities.

The majority of the daily complaints reported to the health department concerned overflowing privy vaults, which remained in use in Milwaukee well into the twentieth century.[64] To empty their vaults, householders hired scavengers, who in turn sold the human excrement as fertilizer to farmers. Many people never bothered to clean their privies, and the "night soil," which Johnson believed "the most dangerous nuisance in the city," spilled over to contaminate the ground and nearby wells.[65] The health department found foul privies on almost every occupied block in Milwaukee, many of them poorly constructed, without cement floors and water-tight walls.

Milwaukee licensed and regulated night scavengers—so called because they made their rounds under cover of darkness—but the city did not require that citizens use them. Many Milwaukeeans never hired the scavengers until forced to do so by the health officer after neighbors reported their negligence.[66] The health department did not initiate any controls beyond nuisance abatement, believing that the removal of night soil was "essentially a private matter," one that should not be usurped by public officials.[67]

U.O.B. Wingate sought to bring privy vaults under public control when he took office in 1890. He found over 20,000 vaults still in use, many located on blocks where sewer and

[63] MHD, *Annual Report*, 1896, p. 22.

[64] MHD, *Annual Report*, 1871, p. 26; MHD, *Annual Report*, 1873, p. 13; MHD, *Annual Report*, 1874, pp. 51-53.

[65] MHD, *Annual Report*, 1878, p. 244.

[66] *Sentinel*, September 2, 1879, p. 8; August 25, 1880, p. 3; August 1, 1884, p. 3. For the number of privies cleaned during the years 1881-1891, see the MHD, *Annual Report*, 1892, table following page 87. Compulsory cleaning of privies increased during the ten years 1881-1891 from 353 in 1881 to 2,797 in 1891.

[67] *Sentinel*, June 29, 1887.

Figure 7. Malodorous privy in Milwaukee, July 1931.
Courtesy of City of Milwaukee Health Department.

water pipes existed and the majority "merely holes in the ground."[68] He proposed to limit vaults to those parts of the city which did not have easy access to water pipes, and to require watertight containers. The public greeted Wingate's ordinance with suspicion. A group of citizens voiced the common sentiment: "[I]t is nothing but a scheme to increase the work of plumbers and masons at the expense of the . . . public."[69] Despite the public outcry, the council passed the regulation, at least in part because it feared an approaching cholera epidemic.[70] "This is one of the most important ordinances, from a sanitary standpoint, that has been passed in the city for years," Wingate proudly proclaimed.[71] By 1910 many Milwaukeeans still used privy

[68] MHD, *Annual Report*, 1891, p. 29.
[69] Letter to the editor, signed "Citizens of the Sixth and Thirteenth Wards," *Journal*, May 26, 1891, p. 4.
[70] MHD, *Annual Report*, 1892, pp. 48-49.
[71] MHD, *Annual Report*, 1893, pp. 10-11.

vaults—in that year the health department ordered 1,000 to be cleaned and another 1,000 removed—but the 1892 ordinance and its 1900 successor decreased the health dangers from human excrement.[72]

Another component of the health department's fight against disease was the collection of mortality and morbidity statistics. The health department, trying to identify the extent of the problem, required Milwaukee physicians to report cases of contagious diseases; yet, because it depended on physician good will, the department rarely prosecuted doctors who did not comply with the regulations. Using the physician reports and mortality data, the health officer compiled monthly and yearly reports on the condition of Milwaukee's health. When a particular disease increased in incidence or an epidemic threatened, the department expanded its efforts. When scarlet fever prevailed in 1878, Dr. Johnson placarded all homes in which the disease had been reported, excluded afflicted children from the public

Figure 8. Travelling disinfectant van, Milwaukee, c. 1910. Courtesy of City of Milwaukee Health Department.

[72] MHD, *Annual Report*, 1910, p. 90.

Figure 9. Street cleaning in Milwaukee, c. 1900. Courtesy of Milwaukee Public Library.

schools, and advertised methods of treatment in city newspapers. Similarly, during an outbreak of diphtheria he outlawed public funerals and distributed Polish, German, and English circulars on how to treat the disease.[73] For both scarlet fever and diphtheria, he advised isolating the patient, boiling or burning all personal effects, keeping the sick room warm and well ventilated, and using disinfectant liberally.[74]

Health officers sometimes clashed with residents over policies aimed at controlling the spread of disease. During the 1891 outbreak of diphtheria, Health Commissioner Wingate tried to keep infected homes quarantined and

[73] MHD, *Annual Report,* 1878, pp. 194, 198.

[74] Physicians debated the extent of isolation necessary for diphtheria patients and in 1891 generally accepted that the longer the isolation the more effective it was. See the Minutes of the Milwaukee Medical Society, December 22, 1891.

66

banned public funerals. Yet despite his orders—which he thought moderate and reasonable—people "persist[ed] in visiting their neighbors thus afflicted, and taking their children along with them." Wingate expressed impatience with those people who, "possess[ing] no realization of the danger of contagion," refused to change their long-held customs at his command.[75] He did not understand them, linguistically or culturally. Walter Kempster also had little patience with people who did not follow his orders when he tried to curb the effects of smallpox in 1894. The educational and ethnic differences between health commissioners and the groups whose life patterns they tried to alter frequently hampered cooperation during times of medical crisis.

Collecting vital statistics led health officers to analyze disease patterns. During the outbreak of diphtheria in 1892, for example, Dr. Wingate identified those wards in the city in which the disease raged, studied the housing, and observed the sanitary condition of the area. He concluded that diphtheria was most prevalent in those parts of the city with the poorest drainage.[76] Since the last areas to be connected to the sewer system and water supply were generally outlying immigrant communities, health officials frequently connected foreign-born areas with high mortality, noting not only their lack of facilities, but also immigrant refusal to cooperate with health department directives.

In addition to its sanitation activities and its collection of vital statistics, the health department also became involved in patient care in hospitals and dispensaries in the nineteenth century. After 1867 the department continued the city's earlier policy of paying physicians to attend the sick poor in hospitals and in contracting with hospitals for emergency services in times of epidemics. In 1871 the Milwaukee City Dispensary, staffed by volunteers from the Milwaukee Medical Society, began for one hour every day

[75] MHD, *Annual Report*, 1891, p. 22.
[76] MHD, *Annual Report*, 1892, pp. 14-16.

"giving gratuitous medical advice and medicine to the poor who are unable to employ a physician."[77] Although the dispensary, located in the health department office, was open only from 4:00 to 5:00 p.m., sick patients came throughout the day, and many health officers found their other duties interrupted to attend to these illnesses. In 1877, for example, 166 patients sought treatment when the dispensary was not open, and Dr. Johnson stopped his work to attend to them.[78]

Dispensary service convinced some Milwaukee doctors of the need for a city hospital to replace the temporary arrangements with the private hospitals. Johnson told of repeated cases of deformity and debility that went untreated because there was no city "provision for this expenditure."[79] Agitation for a city hospital led the city in 1877 to purchase land in the eleventh ward, among the least populated wards in the city. Filled with new Polish immigrants, the eleventh ward lacked the political clout to fight the presence of a hospital in its midst.

The decision to erect a city hospital for the sick poor created new stresses on the relationship between the common council and the health officer. Dr. Wight spent many hours drawing up architectural plans that provided adequate ventilation and space, only to learn that the aldermen had contracted to build the hospital "not in accordance with the demands of modern sanitary science."[80] Upset at the snub, Wight conceded that the new hospital might serve as a "pesthouse for fifteen or twenty smallpox patients," but could be "used for nothing else."[81] Over the protests of the

[77] MHD, *Annual Report*, 1871, p. 37. Minutes of the Milwaukee Medical Society, January 5, 1871, January 19, 21, 1871. Drs. E. W. Bartlett and Julius Kempe were the first in attendance at the dispensary. See also February 5, April 6, 1871.

[78] MHD, *Annual Report*, 1877, p. 98. See also the *Daily News*, August 28, 1879.

[79] MHD, *Annual Report*, 1877, pp. 15-16.

[80] MHD, *Annual Report*, 1878, p. 210.   [81] *Ibid.*, p. 213.

health commissioner, the city went ahead and built a limited-use facility. The new hospital, opened in 1879, had no water or sewer connections and "inadequate" heating arrangements. But Milwaukee had its city hospital, complete with a paraphrased Benjamin Franklin inscription above its door: "Public Health is Wealth."

Wight's predictions about the hospital's limitations proved essentially correct. Because of its physical deficiencies, the hospital rarely admitted patients. In 1882, when Health Commissioner Martin visited the building, he reported that "no patients [were] confined there."[82] In 1890 Wingate noted that the hospital was empty, in bad repair, and ill-regarded among its eleventh-ward neighbors. In keeping with his orderly administration, Wingate worked to upgrade the institution and its reputation. He officially renamed it the "Isolation Hospital" in an attempt to divorce it in peoples' minds from a pesthouse. He built a disinfecting chamber, added water and sewers, and improved the heating system so that the building could be occupied during the winter months. By 1893 the city hospital stood ready to receive patients. In Wingate's opinion, Milwaukee finally had "a model hospital . . . for the treatment of contagious diseases."[83] He encouraged his fellow doctors to use the improved facility and tried to win public acceptance for the remodeled building. But neighborhood fears did not diminish. The public's negative opinion of the hospital became acutely evident during the 1894 smallpox epidemic, when residents forcibly resisted sending their sick to the institution.

Despite such ill feeling, the health department remained in the hospital business. After 1892 it administered the Johnston Emergency Hospital for accident cases and non-

---

[82] *Sentinel*, September 12, 1882, p. 6.

[83] Dr. Wingate gave an address to the Milwaukee Medical Society on the topic of the city isolation hospital, March 28, 1893; typescript attached to the minutes of that meeting. The quote is from p. 12.

contagious illnesses.[84] In 1901 Commissioner Schulz opened Isolation Hospital #2 in a converted old building and put most infectious cases there, leaving the original hospital for smallpox only.

By the turn of the century most physicians and patients at both city isolation hospitals agreed that Milwaukee needed a new building. When the city building inspector closed the second isolation hospital as a fire trap in 1906, the situation grew critical. With outside support from the Milwaukee Medical Society, the health department tried to break down aldermanic resistance and build a new hospital.[85] Finally in 1911 the Socialist common council appropriated the necessary money, and a new wing for the isolation hospital opened in 1912. In 1916 the city added a second wing, and the old hospital was razed. The new building, named South View Hospital to separate it in the public's mind from the previous institution, remains today as a link in the city's health delivery system.[86]

Most health department activity in the late nineteenth century focused on the isolation and prevention of infectious diseases and on sanitation—slaughter houses, water, ice, sewage, privies—because the physicians who held the public positions believed that rotting organic wastes in

[84] Section 7 of the ordinance establishing the Emergency Hospital, passed November 14, 1892, in MHD, *Annual Report*, 1894 and 1895, p. 105.

[85] MHD, *Annual Report*, 1903, p. 33. Milwaukee Medical Society Minutes, May 22, 1906.

[86] MHD, *Annual Report*, 1911, pp. 14-17; Milwaukee Medical Society Minutes, March 14, 1911. For a brief history of the Milwaukee hospital situation and a description of South View Hospital, see "A History of the Isolation Hospital" written ca. 1959, typescript in the Milwaukee Health Department Library, folder 110. The 1910-1912 Socialist city government added other medical facilities to expand the public hospital network. The city took over the Blue Mound Sanatorium for tuberculosis patients in 1911 and greatly increased outpatient facilities available in the city. Children's clinics and adult dispensaries opened in street-front locations in neighborhoods around the city beginning in 1911. See the MHD, *Annual Report*, 1911 and the MHD, *Annual Report*, 1913.

crowded urban areas produced a miasmatic atmosphere conducive to the spread of diseases such as cholera, yellow fever, diphtheria, or typhoid fever. Cleaning the city streets of their most offensive litter would, according to this theory, alleviate the dangers and protect the city against the ravages of disease. By the end of the century, however, regular physicians replaced this miasmatic theory of disease with the germ theory of disease. Instead of blaming dirt for many diseases, physicians increasingly accepted the role of specific germs in propagating specific diseases, basing their changing ideas on the investigations of such European scientists as Louis Pasteur and Robert Koch. The isolation of specific bacteria did not immediately change daily health department work, because application of the new knowledge was not immediately evident. Yet intellectual acceptance among Milwaukee regulars of the germ theory paved the way for practical changes in health department activity, and by 1894, when diphtheria antitoxin became available, health departments moved quickly to accept it. Antitoxin was first used by a private physician in Milwaukee, but very soon after Dr. A. J. Scott obtained his supply from the Pasteur Institute in New York, the health department also secured some antitoxin and quickly established free diphtheria stations around the city.[87] Dr. Kempster attributed the reduced death rate from diphtheria (37.8 percent of cases in 1894; 30 percent in 1895; 21.5 percent in 1896) to this city distribution.[88] The city also used drugstores throughout the city as depositories for culture tubes, analyzed daily in the health department laboratories.[89]

Having spent so much of their time identifying health problems and trying to alleviate them through sanitary cleanups, health departments now changed their emphasis

[87] *Sentinel*, December 6, 1894. *Sentinel*, December 18, 1894. MHD, *Annual Report*, 1895, pp. 8-9. MHD, *Annual Report*, 1895, p. 19.
[88] MHD, *Annual Report*, 1896, pp. 41-42; MHD, *Annual Report*, 1897, p. 45.
[89] MHD, *Annual Report*, 1898, p. 21.

from the general to the specific, although they retained many nineteenth-century practices. Instead of cleaning dirty environments to conquer general health problems, public physicians increasingly sought out specific germs to alleviate specific diseases. The movement to introduce medical services into the public schools illustrates the changing trends. In the 1870s, concerned with the sanitation of schools because of the possibility of epidemic diseases, Health Commissioner Wight personally visited every public school in the city to inspect ventilation, light, heat, privies, and water closets. Shocked by what he saw, Wight began a vigorous and eloquent effort to improve conditions under which students spent their days. He urged aldermen to take a "sanitary primer" in their hands and sit for a week in a schoolroom, where he felt sure they would begin to feel the "headache and depression" that accompanied the close environment. "The habit of dullness, begotten by the unsanitary conditions," Wight warned, "lasts during life, and more or less cripples the productive energy of a whole generation of citizens."[90] The council members heard Wight's evidence and plea, but refused to visit the schools or to allocate any additional funds for their inspection.[91]

Commissioner Wingate reopened the issue of school inspection because parents complained to him that their children suffered from a school-caused epidemic of "sore eyes." Wingate, appalled by the conditions he witnessed in the classrooms, found a few of them "wholly unfit" for school use.[92] His petition to the common council about the schools went unheeded, at least in part because he wanted to in-

[90] MHD, *Annual Report*, 1878, pp. 156-157, 158. The entire report is pp. 129-158.
[91] *Daily News*, February 4, 1880. For the situation in other cities, see John Duffy, "School Buildings and the Health of American School Children in the Nineteenth Century," in Charles E. Rosenberg, ed., *Healing and History: Essays for George Rosen* (New York: Science History Pub., 1979), pp. 161-178.
[92] MHD, *Annual Report*, 1891, p. 15.

spect children as well as physical plants. The health commissioner realized that, although the unsanitary conditions were important, individual child inspections were more essential for effective control. Wingate, and Kempster after him, learned, however, that it was easier to approach the school health problem through traditional nuisance abatement.[93] They alleviated the worst of the privy problems and effected some sanitary improvements through these conventional powers, but without council approval and money they could not initiate medical inspection of schoolchildren or maintain significant control over school health. The germ theory had increased physicians' sophistication about how best to attack diseases in schools, but until they could convince the politicians that pupil inspection was as important as privy inspection, the health commissioners had to continue in traditional channels.

Commissioner Schulz used impending diphtheria and scarlet fever epidemics in 1899 to begin "a perfect system of daily medical inspection" under his emergency powers. He hired additional health inspectors and sent them into the public schools to examine children and to remove the sick ones.[94] His system worked well, but when the disease crisis passed and Schulz approached the council for authorization to continue his program, the aldermen refused to allow the extra expenditures.

School medical inspection slowly gained supporters outside the council. The school board and the Woman's School Alliance sympathized with the necessity of incorporating child medical inspections in health department work. The Milwaukee Medical Society also understood the importance of student inspections and finally rescued the situation for the stymied health officers. In 1906, noting the previous

[93] MHD, *Annual Report*, 1892, p. 12; MHD, *Annual Report*, 1893, pp. 28-30; MHD, *Annual Report*, 1894, p. 39; *Common Council Proceedings*, February 20, 1893, pp. 652-654; March 20, 1893, pp. 735-738; MHD, *Annual Report*, 1896, pp. 37-41. See also the *Sentinel*, June 20, 1896.

[94] MHD, *Annual Report*, 1899, p. 32.

unsuccessful and "meagre attempts" at school medical inspection, the physicians arranged with the school board for a demonstration project. One month of inspection uncovered significant medical problems in school children: pediculosis (lice infestation), diphtheria, and numerous respiratory ailments.[95] The trial suggested that inspections of children could stem the spread of some diseases by identifying infected children and removing them from the schools. More important, it proved to the physicians that such inspections would not threaten their normal practice of medicine. In fact, they had the potential for adding new patients to their practices.

As a result of the experiment, the school board hired its own physicians and instituted daily medical inspection in 1909. In 1911 the health department began examining parochial school children.[96] In 1915 it opened "health advice stations" in the schools.[97] Although Dr. Ruhland assured private physicians that the health department would not give treatment except after ascertaining that patients could not afford to seek private aid, the medical society refused to condone the public expansion into advice stations, which it considered private territory. Blasting health

[95] Minutes of the Milwaukee Medical Society, June 27, 1905. Dr. Akerly brought the subject up for discussion. See also February 27, March 13, and October 9, 1906; November 24, 1908; "Medical Inspection in Milwaukee Schools," *Charities* 18 (1907): 112-113; and the "Annual Report of the Committee of the Milwaukee Medical Society on Medical Inspection of Schools" appended to the MMS Minutes of 1910. Since the annual report deals with events of 1906 and 1907, its appendage to the 1910 Minutes seems incorrect. See also the Minutes of the Milwaukee Medical Society, February 26, 1907.

[96] MHD, *Annual Report*, 1911, pp. 114-115. This inspection was carried out with the cooperation of the Milwaukee Medical Society until 1915. See their *Annual Report*, 1912. See also City Club of Milwaukee Committee on Public Health, *Medical Inspection in the Schools of Milwaukee* (Milwaukee, 1919).

[97] See the Milwaukee *Leader*, June 1915; *Daily News*, May 13, 1915; August 28, 1915; *Sentinel*, February 27, 1919; *Journal*, October 24, 1937; and the clippings box, Local History Room, Milwaukee Public Library.

department encroachment, Dr. Horace Manchester Brown voiced the economic fears of his profession when he lamented that "the young doctor who was unfortunate enough to study medicine hasn't a chance anymore."[98]

School medical inspection epitomized the dilemmas, controversies, and successes of early-twentieth-century health department activity. Secure and expanding in many areas, health officers remained bound by what the local politicians would accept. The traditional nuisance abatement powers allowed physicians a certain amount of leeway, but not enough to wage a full-scale battle with a severe health problem. New medical theory provided new directions for public health activity, but was not sufficient alone to bring about the necessary programs. Successful school medical inspections came only when a coalition of physicians, volunteers, and the school board united with health department workers to convince the political leaders that change was in the city's interest. Although the private physicians turned against increased school activity when it seemed to interfere with their own interests, the temporary alliance had allowed the health department to expand its focus.

Health department medical officers increasingly found themselves at the center of a nexus of people who supported specific changes in health policy. Through this coalition of interest groups, aided by increasingly sophisticated medical theories and technologies, public physicians became more successful at weaving the fabric of health reform. In order to understand how and why the delicate balances between public and private interests and between politicians and physicians produced change, we will examine in detail three public health problems and their resolutions in Milwaukee, each of which represented a major nineteenth-century health concern: smallpox (infectious disease), garbage (sanitation), and milk (food control).

---

[98] *Sentinel*, November 25, 1920. For more on this controversy between the health department and the medical society in the 1920s, see "Socialized Medicine" clippings box, Local History Room, Milwaukee Public Library.

CHAPTER THREE

The Politics of Health Reform: Smallpox

Health Commissioner Walter Kempster took time from his efforts combatting a raging smallpox epidemic in Milwaukee in 1894 to reflect on the public's fearful reactions to this disease. "[T]he alarm caused by a few cases of small-pox," he noticed, "has served to unbalance the equanimity of the entire community." Kempster realized that smallpox, the nineteenth-century "scourge" of Milwaukee, disrupted daily life more than any other disease. Smallpox was physically repulsive, highly infectious, and often fatal. But Milwaukee's smallpox terror emanated more from fear of the unfamiliar than from the physical dangers of the disease. Kempster continued: "[W]here smallpox claims one victim, diphtheria claims its hundreds; still, about the one disease the community becomes frenzied with fear, while about the other little or no attention is given it."[1] Milwaukeeans accepted as unavoidable the ravages of the familiar diphtheria, as they did tuberculosis, scarlet fever, and typhoid fever, and developed ways of coping with these killers. They found it more difficult to adjust to a disease that struck the city only occasionally.[2]

[1] MHD, *Annual Report*, 1895, p. 20.
[2] Cholera would have produced the same fears, but after the 1849-1853 pandemic that dread disease did not visit Milwaukee. Yellow fever and cholera produced similar reactions in cities unfamiliar with them. Charles Rosenberg, studying the effects of cholera epidemics on American cities,

Medical disagreements about smallpox prevention and treatment compounded the public's discomfort. Some physicians advocated vaccination as a sure protection against smallpox; others proclaimed vaccination more dangerous than the disease it sought to prevent.[3] Some physicians believed that smallpox victims should be treated in an iso-

noted: "the cholera epidemics of the nineteenth century provided much of the impetus needed to overcome centuries of governmental inertia and indifference in regard to problems of public health. . . . It is not surprising that the growing public health movement found in cholera an effective ally." *The Cholera Years: The United States in 1832, 1849, and 1866* (Chicago: Univ. of Chicago Press, 1962), pp. 2-3.

See also John Duffy, who posits that cholera and yellow fever were "important factors in promoting public health measures," because of their "crisis" presentation. "Social Impact of Disease in the Late Nineteenth Century," *Bulletin of the New York Academy of Medicine*, 47 (1971): 800. Although Duffy saw smallpox running a poor third to cholera and yellow fever, in Milwaukee smallpox took the place of the former two diseases, which did not threaten the city after 1850. I make the parallel with cholera and yellow fever despite the major differences between those diseases and smallpox, the availability of a preventive for one and not for the others. While vaccination raises interesting differences between the examples used here, those differences did not affect the public reaction evoked in each case: fear and panic and an immediate governmental response to alleviate conditions. Typically, mortality rates in the late-nineteenth-century smallpox epidemics stayed under 30 percent, although they could have been as high as 50 percent. See M. V. Ball, "Deathrate from Smallpox in Various Cities and States," *American Medicine* 5 (1903): 450.

[3] See Martin Kaufman, "The American Anti-vaccinationists and Their Arguments," *Bulletin of the History of Medicine* 41 (1967): 463-478. Kaufman describes most anti-vaccinationists as irregular practitioners and identifies the movement with sectarian medicine. It is clear to me that the division does not hold in Milwaukee, where many regularly trained physicians were hesitant about the protective value of vaccination and where many sectarians supported vaccination. Also, the distinction between regular and irregular physicians was a foggy one in the minds of most people who sought medical advice and therefore it is not particularly useful for understanding the acceptance of anti-vaccinationist thought in nineteenth-century American cities. See also Elizabeth Barnaby Keeney, Susan Eyrich Lederer, and Edmond P. Minihan, "Sectarians and Scientists: Alternatives to Orthodox Medicine," in Ronald L. Numbers and Judith Walzer Leavitt, eds., *Wisconsin Medicine: Historical Essays* (Madison: University of Wisconsin Press, 1981).

lation hospital; others thought home treatment most beneficial. Political and ethnic divisions in the city often exacerbated the medical confusions as coalitions formed around the conflicting ideologies. Typically, the health department used the lack of consensus in the city to step in and increase its authority to control infectious diseases. During only one epidemic, in 1894-1895, did the confusions of the moment lead to a restriction, rather than expansion, of health department powers. Milwaukee's response to smallpox is important not because that disease killed so many people, but because smallpox prodded the city into developing its pattern of control over many diseases. Conquering smallpox did not significantly improve Milwaukee's overall health; it did greatly facilitate the growth of health department authority.

Before the city created a permanent board of health, smallpox struck Milwaukee twice in 1843 and in 1846. In both instances the common council hired physicians to serve as a board of health and gave them authority "to take the necessary steps" to deal with the emergency. The board rented space from the Sisters of Charity for a pesthouse, where it treated sick patients who could not be cared for elsewhere. It requested that physicians report all their smallpox cases and recommended vaccination. When the epidemics abated, the boards of health disbanded.[4]

The city willingly accepted this limited responsibility for emergencies caused by periodic epidemics. Officials delegated money and authority when disease struck the city and provided aid until the disease abated. They made no move toward preventing future disasters. When the per-

[4] For accounts of smallpox in Milwaukee in the 1840s, see Walter Kempster, "History of Smallpox in Milwaukee," in J. Watrous, ed., *Memoirs of Milwaukee County* (Milwaukee, 1909). The version used here is from that article, reprinted in Frank, pp. 178-180. See also Peter Harstad, "Disease and Sickness on the Wisconsin Frontier: Smallpox and Other Diseases," *Wisconsin Magazine of History* 43 (Summer, 1960): 256-258; and the MHD, *Annual Report*, 1871, pp. 7-8.

manent board of health began operations in 1867, a broader perspective on urban diseases became possible, as was immediately evident in Dr. James Johnson's approach to his first emergency, the smallpox epidemic of 1868.

Physicians reported 501 cases of smallpox to the new health office between December 1868 and April 1869: 419 cases (84 percent) were among Germans, who constituted only one-third of Milwaukee's population.[5] At Dr. Johnson's request, the common council contracted with the Passavant Hospital to accept smallpox patients, although Johnson made no attempt to force patients into the hospital. During the four-month epidemic the facilities were "taxed to [their] utmost capacity" with 123 public charges.[6] In addition to providing care for the sick, Johnson tried to protect the well by closing schools in the German west side, where smallpox raged most virulently.[7]

Although Johnson admitted that vaccination was "not always successful," he proclaimed unequivocally that it was "the only known preventive" that could "save the city from the ravages of the smallpox."[8] A vigorous vaccination policy became the pivot of Johnson's campaign to eradicate smallpox. Using his emergency powers, Johnson appointed six physicians to visit schools and homes to vaccinate residents free of charge.[9] He wrote letters to the newspapers lauding the protective powers of vaccination and urging citizens to vaccinate their children.[10] Sixteen hundred Milwaukeeans took advantage of the free service, while untold numbers

[5] MHD, *Annual Report*, 1869. W. C. Bennett, "Smallpox and Vaccination in Wisconsin," *Transactions of the State Medical Society of Wisconsin*, 1902, pp. 138-139; Kempster, pp. 178-179.

[6] *Sentinel*, June 15, 1868, July 28, 1868; MHD, *Annual Report*, 1871, pp. 12-13. The Passavant was also known as Milwaukee Hospital. MHD, *Annual Report*, 1868, p. 40.

[7] MHD, *Annual Report*, 1871, p. 13; *Sentinel*, November 21, 1868; December 5, 1868, p. 1.

[8] *Sentinel*, April 13, 1870.

[9] *Sentinel*, November 25, 1868, p. 1; MHD, *Annual Report*, 1869, p. 42.

[10] See, for example, *Sentinel*, November 28, 1868, p. 1.

of other residents rejected Johnson's therapy—as well as his authority to dispense it. For his efforts the health officer found himself the target of lively controversy in the daily press.

Johnson provoked negative reactions by blaming "a certain class of medical practitioners" for the concentration of smallpox in "certain [German] wards of our city."[11] He linked his vaccination opponents with the sectarians, even though some homeopaths were among his most ardent vaccination supporters. Instead of offending the sectarians, Johnson's accusations brought an immediate rebuke from a group of regular physicians, led by prominent Dr. Ernst Kramer, who repudiated vaccination. Kramer found Johnson's press statement "arrogantly and insultingly expressed," an attempt "to deliver [Kramer and his colleagues] up to the hate and contempt of a frightened community, by holding them responsible for . . . all those cases which proved fatal." Implying that he had the support of the Milwaukee City Medical Association, and pointing out that "some of the very highest authorities . . . put a *very low* estimate upon the value of vaccination," Kramer proclaimed that "we do not recognize [Johnson's] self-constituted authority and much prefer . . . to use and act according to our own judgment in scientific matters."[12]

Kramer represented a growing group of anti-vaccinationists in Milwaukee, many of whom were German, who feared that the dangers of vaccination outweighed its possible benefits. Both sectarian and regular physicians supported this position. Their concern emanated from the vaccination procedure itself and the variations in its practice. Some practitioners still used "humanized" virus, matter taken from the pustules of inoculated humans, which carried the added danger of infecting recipients with such

[11] *Evening Wisconsin*, January 16, 1869, quoted in the *Sentinel*, January 20, 1869.
[12] Letter to the editor, *Sentinel*, January 22, 27, 1869; for more on Kramer, see Frank, p. 44.

diseases as syphilis. If transmitted through too many people, the virus lost its immunity powers. Most American vaccinators by the last third of the nineteenth century used bovine matter, which they considered safer as well as more effective, but which, when kept too long, also gave minimal protection against smallpox. Furthermore, the inoculation procedure could be a cause for worry. Vaccinators, sectarian and regular physicians as well as midwives and informal practitioners, used scarification, puncture, or abrasion to prepare the skin for the introduction of the viral matter from the quill or ivory "point" on which it was stored. Vaccinators made from one to sixteen punctures, usually on the arms, which, even under ideal conditions, became badly scarred. Some vaccinators exercised great care, cleaning the skin and their lancets and making precise incisions; others, untrained or in a hurry, used a single point and lancet for many people and failed to cleanse the puncture site. The variation in the quality of the vaccine matter and in inoculation practices led to enormous variation in outcome, and reports of negative effects abounded. Many concluded that people were safer without vaccination than with it. Because of the professional doubts about the efficacy and safety of vaccinations, lay people could readily agree with Kramer's conclusions that "the books are not closed yet on the subject of vaccination." In the face of uncertainty it seemed wise to wait.[13]

Despite the "effusion of venomous abuse" in the letters to the newspapers, Johnson continued his avid support of vaccination. He, like Kramer, claimed that the medical society supported his policy and that "leading German doctors as well as all the English speaking physicians, allopathic

---

[13] *Sentinel*, January 27, 1869, p. 1. For a contemporary account of the vaccination procedure, see W. A. Hardaway, *Essentials of Vaccination; a Compilation of Facts Relating to Vaccine Inoculation and Its Influence in the Prevention of Smallpox* (Chicago: Jansen, McClury & Company, 1882). See also a lengthy discussion of nineteenth-century practices in Cyril W. Dixon, *Smallpox* (London: J & A Churchill Ltd., 1962), pp. 249-295.

and homeopathic," favored vaccinations.[14] The dispute between Johnson and Kramer made it impossible for many Milwaukeeans to decide whether or not to be vaccinated. One confused citizen portrayed the common dilemma: "On the one hand we have been informed by certain physicians that vaccination not only does no good but is really injurious. On the other hand the Board of Health have advised *all* to get vaccinated, and have employed physicians for that purpose. By this disagreement of the doctors a great many people are undecided what to do."[15] The Milwaukee City Medical Association tried to settle the confusions by unanimously supporting Johnson's policy and urging all citizens to get vaccinated by "respectable" physicians.[16] Kramer did not respond to this action, but the issue was far from settled.

Established medical opinion notwithstanding, the school board followed Kramer's precepts. It decided to allow unvaccinated children to attend school after a ninth-ward (German) public-school principal, with the full support of the parents in his community, refused to admit a city physician into his school to vaccinate the students. The school board agreed that pupil attendance at school should not be restricted for a "generally disputed" medical practice: "The admission to the public school should be free to all, in the most extended sense of the term. No obstruction whatever should be interposed."[17]

The resistance to Johnson's vaccination policy emanated largely from the German community. Many German-Americans believed, with Kramer, that vaccinations could

[14] *Sentinel*, January 26, 1869, p. 1.

[15] Letter to the editor, *Sentinel*, January 25, 1869, p. 1.

[16] MCMA, Minutes of regular meeting, February 18, 1869, in the Milwaukee Academy of Medicine. The physicians who either attended the meeting and voted for the resolutions or those who endorsed it later, included Drs. J. K. Bartlett, Alfred Castelman, M. P. Hansen, H. E. Hasse, J. Johnson, S. Marks, L. McKnight, W. Thorndike, E. B. Wolcott, E.H.G. Meachem, J. M. Allen, Jr., R. D. McArthur. See also *Sentinel*, February 22, 1869, p. 1.

[17] *Sentinel*, October 11, 1869, p. 1; October 28, 1869, p. 1.

be dangerous. Some others, having come to America seeking political freedoms, would not support a government policy that seemed to invade their private lives. As one physician came to realize, they wanted "their personal liberty." German refusal to cooperate with Johnson stemmed, at least in part, from this refusal to accept an extension of public authority.[18] Resistance to Johnson grew even stronger when he continued to advocate vaccination after the smallpox epidemic receded in 1869. Council members, sensitive to public sentiment, ignored Johnson's pleas for systematic vaccinations of schoolchildren.[19] Johnson found that he needed the next smallpox epidemic to help him to extend public health activity beyond the emergency itself.

The health officer made some headway in consolidating his authority to control infectious diseases during the 1871-1873 smallpox epidemic. The disease again struck hardest among the unvaccinated Germans: in 1871, 568 (76 percent) of 744 cases were German; in 1872, 469 (76 percent) of 616 cases were German.[20] As soon as smallpox appeared, Johnson activated his controls. He sent hand-delivered circulars to physicians to remind them of their duty to report the disease.[21] He reopened and staffed the Passavant pest-

[18] Dr. Nazum quoted during a discussion of W. C. Bennett, "Smallpox and Vaccination in Wisconsin," *Transactions of the State Medical Society in Wisconsin*, 1902, p. 154. Compulsory vaccination was not always accepted even by those physicians who advocated vaccination. This added division among the medical community did not help Johnson's cause. The medical literature of the day indicated that other cities were having similar problems. See, for example, J. M. Toner, "A Paper on the Propriety and Necessity of Compulsory Vaccination," *Transactions of the American Medical Association* 16 (1865): 307-330; and William H. Richardson, "Smallpox in New York City, with Some Statistics and Remarks on Vaccination," *Transactions of the New York State Medical Society*, 1865, pp. 143-156. The issue of compulsory v. voluntary vaccination became more important later in the century, and will be examined more fully later in this chapter.

[19] *Sentinel*, April 13, 1870, p. 1; July 28, 1871, p. 4; December 4, 1871, p. 4; MHD, *Annual Report*, 1870, pp. 3-5, 1871, pp. 24-26.

[20] MHD, *Annual Report*, 1872, p. 28; MHD, *Annual Report*, 1873, pp. 38-39, 52. See also Kempster, pp. 178-180; and Bennett, pp. 138-139.

[21] *Sentinel*, December 31, 1871.

house, continuing his policy of removing only those who wanted to go, mostly "hired girls, single men without parental homes, strangers in boarding houses," and people with no one to care for them.[22] He published instructions on the care of smallpox patients and left copies at every house that reported illness.[23] He outlawed wakes and public burials and insisted that bedding and clothing of victims be destroyed or thoroughly disinfected.[24] He placarded homes containing smallpox patients.[25]

Vaccination remained the most important part of Johnson's attack on smallpox. Through the well-advertised City Dispensary he offered free vaccinations for everyone.[26] He worked to get the state legislature to require vaccination of schoolchildren, but the measure did not pass.[27] Johnson was more successful on the local level when he convinced the school board in 1873 to reverse its earlier decision and to require vaccination for all incoming students.[28] He believed that his campaign had overcome "most of the stumbling-blocks" to controlling infectious disease.[29] The one blockade that Johnson could not topple was the resistance of the anti-vaccinationists. He stood by helplessly watching smallpox devastate the unvaccinated German population. He again blamed "those physicians and the conductors of newspapers who have denounced vaccination." It was tragic,

[22] *Sentinel,* January 16, 1872, p. 4. Johnson tried to get the city to provide money for its own hospital, but he was unsuccessful. MHD, *Annual Report,* 1872, p. 45.

[23] The instructions were reproduced in the MHD, *Annual Report,* 1872, pp. 41-42. See also, *Sentinel,* January 3, 1872, p. 4.

[24] MHD, *Annual Report,* 1872, p. 43.

[25] *Sentinel,* June 11, 1872, p. 4.

[26] MHD, *Annual Report,* 1872, pp. 39-40; *Sentinel,* January 16, 1872, p. 4.

[27] MHD, *Annual Report,* 1872, p. 44; *Sentinel,* January 26, 1872, p. 2; February 14, 1872, p. 4; January 20, 1873, p. 4.

[28] MHD, *Annual Report,* 1874, pp. 14-15; *Sentinel,* February 20, 1873, p. 4; March 3, 1873, p. 4.

[29] Johnson letter in *Sentinel,* January 20, 1873, p. 4.

he mourned, that the "grave has closed over many of the dupes of their insane councils."[30]

Johnson did not seek compromise with those who differed with him about vaccination. He regarded the anti-vaccinationists as a formidable political enemy to the health office, to be overpowered if possible. Johnson's refusal to attempt conciliation served to harden the opposition against him and confirmed his opponents' fear of unbending governmental authority. The health officer felt the force of resistance even more strongly during the next smallpox epidemic, which struck Milwaukee in 1876 and 1877.[31]

In 1876 and 1877 smallpox again disproportionately attacked the German population. This time they were joined by the newest immigrants to Milwaukee, the Poles, who, fearing governmental authority and the dangers of vaccination, also refused vaccination and suffered even more severely than the Germans.[32] Johnson continued to fight the disease as he had in the past, vaccinating and isolating, but during this epidemic the pesthouse issue, rather than vaccination, highlighted much of the debate in the city.

As it had the previous two epidemics, the common council appropriated money to subsidize the Passavant pesthouse but chose to ignore Johnson's plea for a city hospital. Council members did not see the need for a city-owned hospital. One alderman insisted that there was "no use for a hospital" since most people "would rather attend to those of [their] own family who might be afflicted" than send them to a pesthouse.[33] Perhaps even more to the point, aldermen knew "there was bitter prejudice against the building of a smallpox hospital anywhere in the city" and

[30] MHD, *Annual Report*, 1872, p. 7.

[31] MHD, *Annual Report*, 1876, p. 25; MHD, *Annual Report*, 1877, p. 123.

[32] MHD, *Annual Report*, 1876, p. 57. In 1876, out of 495 reported cases of smallpox, Germans and their children suffered 182 and Poles and their children, 232; together the two groups accounted for 84 percent of the cases.

[33] Alderman Wolf (5th ward) quoted in *Sentinel*, June 7, 1876, p. 2.

did not want to take the political risk of supporting an unpopular idea.[34]

But the council could not continue to ignore the issue of a city hospital. Second-ward residents, who lived near the Passavant pesthouse, complained that "[i]t is not pleasant— to say the least—to be at any time in close proximity to institutions of this character," and they demanded its removal from the center of town.[35] Responding to this citizen pressure, the council declared the rented pesthouse a "public nuisance."[36] The Passavant's board was astonished. It refused to "be sacrificed by the city using its grounds in opposition to public remonstrances" and announced the hospital would accept no more city patients.[37] The *Sentinel* reflected on the irony of the Passavant's position: it accepted smallpox patients at the request of the city, and then the city declared the hospital a nuisance because it had smallpox inmates! "There is a strong flavor of impudence about the action of the Common Council," concluded the wry editorial.[38]

Impudent or not, the aldermen's own action forced them into building a public hospital. The city purchased land in the eleventh ward, where the new Polish immigrants were too sick and too new to Milwaukee politics to protest, and built an imposing structure of local cream-colored bricks. The hospital was ready for occupancy in July 1879, by which time smallpox had disappeared from the city. Its lack of sewer and water connections and its incomplete heating system marked the new hospital as inadequate to meet the city's needs from the very beginning.[39]

[34] *Sentinel*, June 14, 1876, p. 8; *Common Council Proceedings*, June 6, 1876, p. 62.

[35] *Sentinel*, February 6, 1877, p. 3. See also November 22, 1876, p. 3; November 23, 1876, p. 3.

[36] *Sentinel*, February 24, 1877, p. 3; March 13, 1877, p. 3.

[37] Quoted from Dr. Passavant's *Annual Report to the Board of Visitors* in the *Sentinel*, March 23, 1877, p. 3.

[38] *Sentinel*, March 23, 1877, p. 4.

[39] *Evening Wisconsin*, July 24, 1879; U.O.B. Wingate, "City Isolation Hospital," paper delivered at the meeting of the Milwaukee Medical So-

One of the earliest questions raised about the city hospital was whom it should serve. Johnson, having battled two previous epidemics as health officer, felt that patients should be removed to the hospital—forcibly if necessary—if their isolation would help to preserve the health of their families and neighbors. He thought that the patients themselves would also benefit, since only three percent of patients died in the smallpox hospital, whereas mortality had been as high as fifty percent in some parts of the city.[40] But the city attorney did not think it proper or legal to take patients who did not want to go. Thus he ruled: "Every person may claim the right, when sick, to stay in his own house, to be nursed by his relatives, and to be attended by the physician of his own choice."[41] Public opinion sided with the attorney, leaving Johnson with a city hospital but no control over whom he could admit to it. "We fail to see the wisdom of permitting a family to deal out death and hideous deformity to a whole community," he lamented.[42] He saw one side of the issue, the public's health, while many people and politicians saw another side, their individual rights. For a while it seemed as if the standoff produced by the two incompatible positions might permanently halt the advancement of public controls over infectious diseases.

But Johnson won more than he lost in his battle for control over infectious diseases during the 1876-1877 smallpox epidemic. The council passed a placard ordinance that provided for vivid labeling of houses sheltering people suffering from smallpox, scarlet fever, or diphtheria. Some people protested that the ordinance invaded their privacy, others that it would hinder commerce; but placarding was

ciety, March 28, 1893, typescript in "Scientific Proceedings" of the society at the Milwaukee Academy of Medicine. See also "A History of the Isolation Hospital," typescript in the Milwaukee Health Department Library, n.d.

[40] MHD, *Annual Report*, 1876, p. 47.

[41] Quoted from Emil Wallbur's decision, in the MHD, *Annual Report*, 1877, pp. 48-49. See also, *Daily News*, August 7, 1894.

[42] MHD, *Annual Report*, 1876, pp. 49-50. See also letter to the editor of the *Sentinel*, November 1, 1877, p. 2.

more of an amusing diversion than a divisive issue in the municipal debate over smallpox.[43]

More serious was vaccination. Johnson hired extra physicians to vaccinate residents, as he had in the past, and again his emissaries were frequently repelled when they tried to carry out their duties. The German press kept emphasizing the dangers of vaccinating, bending only to concede that five out of every one hundred persons vaccinated might benefit from its protection.[44] The new German and Polish immigrants in the south-side wards joined the older west-side Germans in resisting the intrusion of health officials. Many of the newer residents never sought medical aid or reported the disease, making control over its spread virtually impossible. Johnson even sent a Polish police officer to make a "house to house examination" to locate the sick, with a Polish homeopathic physician to vaccinate them, but most cases still went undetected. Johnson mobilized the south-side churches for their "hearty co-operation and best assistance in aid of the good work" of the board of health. But he was not optimistic that smallpox could be controlled in the Polish areas. "So little fear have this class of people of Small-pox," he observed, "that they will eat and drink, cook and sleep in the same apartment with one or more Small-pox patients."[45] The smallpox eradication campaign received its biggest boost when the

[43] Alderman Wolf suggested that *doctors* wear placards as they went through the city; others thought houses *without* disease should be placarded. *Common Council Proceedings*, November 12, 1877; MHD, *Annual Report*, 1877, p. 100; *Sentinel*, November 13, 1877, pp. 2-4. See also *Sentinel*, November 5, 1877, p. 4; February 2, 1877, p. 4; January 31, 1877, p. 8.

[44] German newspaper quoted in the *Sentinel*, November 27, 1876, p. 8. The health office was sensitive to the charge of harmful vaccinations, and made public its exclusive use of "non-humanized virus" claiming that no "unpleasant" effects had resulted from vaccinations done through the health officials. MHD, *Annual Report*, 1877, p. 51. For more on opposition to vaccination, see *Sentinel*, November 20, 1876, p. 4; May 28, 1877, p. 2.

[45] MHD, *Annual Report*, 1877, pp. 42, 55, 43.

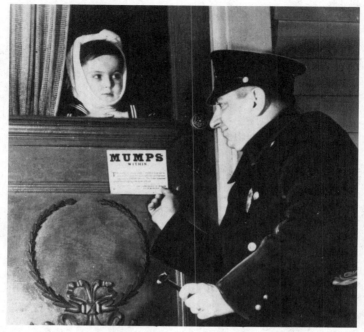

Figure 10. Sanitary officer placarding a home with "MUMPS WITHIN" sign, Milwaukee, 1922. Courtesy of City of Milwaukee Health Department.

common council passed an ordinance upholding the school board's earlier resolution requiring vaccination before a child could be admitted to the public schools.[46]

Johnson's success in building a city hospital, placarding residences, and vaccinating schoolchildren resulted largely from the changing complexion of city life. The 1870s wit-

---

[46] *Common Council Proceedings,* July 9, 1877, pp. 97-98; August 13, 1877, p. 131. See also *Annual Report of the School Board of the City of Milwaukee for the Year Ending August 31, 1877.* Rules and Regulations of the School Board, Article XII, p. xxix, read: "No child shall be received, for the first time, into any public school unless the parent or guardian shall furnish to the Principal satisfactory evidence that said child is not less than four years of age, and has been vaccinated."

nessed the first major influx of poor central European immigrants to Milwaukee. As large numbers of Poles and Bohemians filled the city, the established groups, feeling their way of life threatened, developed anti-immigrant sentiments. Because smallpox concentrated in the south side, where most of the newer, nonimmune, and unvaccinated immigrants lived, settled Milwaukeeans blamed them for its presence.[47] Business and professional people increasingly worked together to support board of health policies that primarily restricted the behavior of the newer immigrant laborers, most severely hit by the epidemic. The class and ethnic divisions over health policy during the 1876-1877 smallpox epidemic were striking. "The best people of the city will stand by the Board in any effort it makes to stamp out smallpox," noted a *Sentinel* editor.[48] The eagerness of the established citizens stemmed at least in part from a motivation to protect their own lives and interests from "the careless and slovenly foreigner": "The Polack element is palpably and culpably careless as regards this disease, and as the class is sufficiently numerous to keep the evil afloat perpetually, the representative portions of the city, who are the real sufferers, begin to find that they have a duty to perform in enforcing the best sanitary measures."[49] The "duty" became even more pressing when Milwaukeeans realized that smallpox cost the city dollars "beyond" calculation.[50]

[47] *Sentinel*, October 30, 1876, p. 8; November 2, 1876, p. 8.

[48] *Sentinel*, editorial, November 5, 1877. Businessmen and physicians together petitioned the common council in support of the placard ordinance, *Sentinel*, November 6, 1877.

[49] *Sentinel*, November 2, 1877, p. 2.

[50] The health officer, interviewed by the *Sentinel*, November 3, 1877, p. 3. See also *Sentinel*, November 21, 1876, p. 3. Citizens debated the question of whether hiding an epidemic or openly dealing with it was better for business. One citizen felt more "howling" should be heard—that alarm would work for the general health of the community rather than against it. "Neighbor" writing letter to the editor, *Sentinel*, July 16, 1877, p. 3. See negative response to that letter in *Sentinel*, July 19, 1877, p. 3.

Johnson won victories in consolidating the board of health's ability to combat infectious diseases largely because of the developing coalition of people economically and socially motivated to support him. He left the health office in 1877 confident that he had established important policy precedents and that the future of public health activity in Milwaukee was secure.

During the next decade, events did follow in patterns that Johnson had established. The health department successfully turned back the threat of smallpox in 1881 and 1882 with vigorous action and community support. Commissioner Wight kept constant vigilance over the railway stations and steamboat landings, hounded the doctors into reporting every case they saw, and advertised any cases that appeared.[51] Placarding was especially important to Wight because, as he said, "my experience convinced me that a community will give wide berth to smallpox . . . if you will only let them know where it is." He had no patience with those who thought placarding an infringement of personal liberty, a liberty he thought "had better be abridged."[52] Continuing Johnson's traditions, Wight outlawed public funerals, advocated disinfection of personal belongings, circulated directions for hygienic management of the disease-stricken, and loudly proclaimed the virtues of vaccination. Wight, showing his ministerial heritage, disagreed with Johnson only about who should use the city hospital: "To force away children from parents . . . for the purpose of isolating infectious disease, exposes the public by transportation of the afflicted, diminishes the chances of recovery, outrages the unreasoning affections, and invades the sanctity of the home. . . . [The] isolation of contagious and

---

[51] *Sentinel*, April 2, 1881, p. 2. See, for example, statements issued to the press in the *Sentinel*, April 14, 1880; January 13, 1881; February 12, 1881, p. 5; March 29, 1881, p. 5; April 4, 1881, p. 3; April 7, 1881, p. 5; May 7, 1881, p. 3; July 13, 1881, p. 2; August 11, 1881, p. 4.

[52] O. W. Wight, "The Management of Contagious and Infectious Diseases in Milwaukee," *Public Health: Reports and Papers* 6 (1880): 90-91.

infectious diseases in the family, which is the unit of our political society and Christian civilization, is wisest and best."[53]

Dr. Robert Martin adhered closely to what he had learned as Wight's assistant. When cases of smallpox appeared in 1882, Martin isolated, disinfected, and vaccinated, and in one case, when the disease appeared in a house across the street from a school, closed the school and burned the afflicted structure "to the ground." This action caused "a great deal of commotion," but Martin felt vindicated when no further cases appeared in the neighborhood.[54]

During the next ten years smallpox rarely appeared in the city, and no Milwaukeeans died from the disease between 1883 and 1893.[55] People began to forget the fears that smallpox generated, and one south-side resident ridiculed the "expensive building" constructed to combat the disease: "[The] only purpose that the building has served is to shelter a janitor and his family."[56]

The neglected and under-used city hospital attracted the attention of Dr. Wingate soon after he became health commissioner in 1890. Eleventh-ward residents tried to eject the pesthouse beyond the city limits into the unpopulated countryside. Wingate countered their activities by mobilizing the city's physicians to support the existing structure.

[53] *Ibid.*, p. 92.

[54] *Sentinel*, April 27, 1882, p. 3. Kempster attributed the burning to Dr. Wight, but he had left the city for Detroit in October 1881, and Martin had been appointed to finish out his term (p. 180).

[55] *Sentinel*, June 29, 1887. Health Department *Monthly Reports*, 1885-1891; MHD, *Annual Report*, 1893. It is interesting, especially in light of 1894 events, that Dr. Martin faced a possible impeachment in 1889 over his "negligent" handling of a smallpox case referred to the city from the county. The common council, blaming Martin for exposing citizens while transporting the patient, considered a resolution to remove him from the position of health commissioner. Testimony did not support the charge of negligence—and, it might be added, smallpox did not gain a foothold in the city. Martin was exonerated. See the *Common Council Proceedings*, June 3, 1889, p. 115; July 1, 1889, p. 185.

[56] *Sentinel*, May 4, 1888.

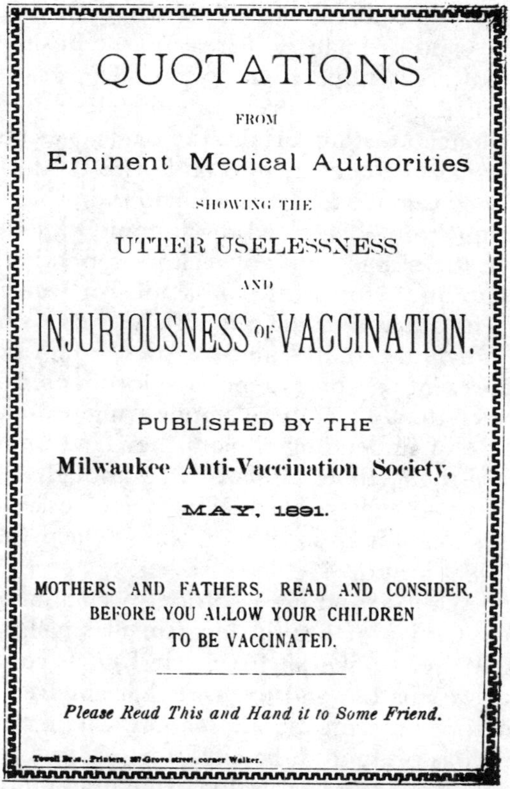

QUOTATIONS

FROM

Eminent Medical Authorities

SHOWING THE

UTTER USELESSNESS

AND

INJURIOUSNESS OF VACCINATION.

PUBLISHED BY THE

Milwaukee Anti-Vaccination Society,

MAY, 1891.

MOTHERS AND FATHERS, READ AND CONSIDER,
BEFORE YOU ALLOW YOUR CHILDREN
TO BE VACCINATED.

Please Read This and Hand it to Some Friend.

Tovell Bros., Printers, 387 Grove street, corner Walker.

Figure 11. Cover of the Milwaukee Anti-Vaccination
Society pamphlet, 1891. Copy in the Middleton Health
Sciences Library, University of Wisconsin-Madison.

Many physicians came to Wingate's aid, petitioning the
common council to follow "intelligent public opinion" and
modernize the original building.[57]

More influential than the doctors, a threat of cholera in
the fall of 1892 brought the council into agreement with

[57] MHD, *Annual Report*, 1891, pp. 7-8; *Sentinel*, October 29, November
1, 3, 5, 1891; MHD, *Annual Report*, 1892, pp. 7-9.

93

Wingate. In their anxiety the aldermen passed a series of resolutions and ordinances increasing the health commissioner's powers to impose quarantine and placard houses; the council also agreed to renovate the eleventh-ward isolation hospital. Most important, the panic over impending cholera induced the council to authorize the health commissioner to remove to the city isolation hospital any sick person who "cannot be cared for at home without especial danger to the public" and any well person "who has been exposed to any dangerous, contagious or infectious disease."[58] The public's health and the city's welfare now came before individual desires and liberties. South-side aldermen, representing constituencies who opposed the pesthouse, tried to obstruct these empowering ordinances, but in the face of impending cholera they could not win the support of their fellow legislators.[59] Although the cholera epidemic never materialized, Wingate remodeled the city hospital, making it an institution about which "Milwaukee can well feel proud."[60]

During Wingate's tenure in office, vaccination remained the most divisive of health department smallpox-related policies. Wingate, like all his predecessors, continued to advocate vaccination and to insist that children be vaccinated before entering public school, whether or not an epidemic was present in the city. In reaction to this policy and to the gains the health department accomplished on other fronts, a group of Milwaukeeans, many of whom were German- and Polish-speaking, established the Milwaukee Anti-Vaccination Society. Their literature urged revocation of the regulation prohibiting school attendance without a vaccination certificate and emphasized that "eminent medical authorities" believed the procedure dangerous.[61] Win-

---

[58] Ordinance is reprinted in the MHD, *Annual Report*, 1893, pp. 14-16.
[59] Wingate, "City Isolation Hospital," p. 27.
[60] MHD, *Annual Report*, 1893, pp. 17-18.
[61] "Quotations from Eminent Medical Authorities Showing the Utter Uselessness and Injuriousness of Vaccination." Published by the Milwau-

gate thought the group "very active" among the "masses of the people" and realized their work hampered the success of the school vaccination program.[62] He sought the aid of the Milwaukee Medical Society to help to combat the new pressure group. During the medical society's debate about whether or not to issue a public statement on vaccination, Dr. Horace Manchester Brown indicated how the medical society viewed its opponents. "The cranks who oppose vaccination belong to that class of incomprehensible blockheads who will not be convinced," he maintained, "and I think it will be quite useless to attempt to educate them."[63] Other physicians nonetheless encouraged the society to take active steps to counteract the anti-vaccinationists' influence, and Wingate ultimately succeeded in getting the physicians' public support for his measures.

On one thing Wingate and the anti-vaccinationists could agree: the danger of bad vaccine administered under non-sterile conditions. Wingate lamented that in Milwaukee "incompetent persons" performed vaccination in a "loose way . . . with the worthless virus that abounds in the market."[64] People could buy their own vaccine points in the drug-stores, and inoculate themselves, or they could purchase the services of any number of variously trained and untrained vaccinators. Because of the uncontrolled possibilities, many of the anti-vaccinationists' accusations about the misfortunes of those who had sought protection could be documented from local experiences. Wingate differed from the anti-vaccinationists, however, about the solution to this intolerable situation. Whereas the latter wanted vaccinations stopped, the health commissioner maintained that absolute

kee Anti-Vaccination Society, May 1891. Pamphlet file, Middleton Medical Library, University of Wisconsin-Madison.

[62] U.O.B. Wingate, "Smallpox in Wisconsin, from January 1894 to June 1895," *Public Health: Reports & Papers of the APHA* 21 (1896), p. 268.

[63] Minutes of the Milwaukee Medical Society, January 12, 1892. See also January 26, 1892.

[64] MHD, *Annual Report*, 1891, p. 17.

public controls over the vaccine matter and its administration, taking it out of "the hands of commercial interests," would ensure safety and effectiveness. Wingate favored strong laws to "compel [people] to do that which is best for themselves and the community in which they live."[65] Thus the lines of contention were drawn in Milwaukee by the time that Wingate left office in 1894: on the one hand stood the health department, increasingly powerful and supported by the community elites; on the other hand stood the anti-vaccinationists, beginning to organize and supported by a varied group of people, many of them immigrants favoring personal independence.[66]

Walter Kempster, who followed Wingate as health commissioner, could not have agreed more with his predecessor that people in positions of authority had to make decisions for the uneducated masses on issues affecting the public's health. Although they belonged to different political parties, Wingate and Kempster were similar in outlook and in medical ideology. Both considered themselves among the elite of the medical profession. Both had high standards for medical practice and for public health. Both had a refined contempt for those who did not share their values of order and cleanliness or who did not accept their authority in office. But their four-year terms as chief medical officer of Milwaukee could not have been more different. Wingate had consolidated and increased the power in the department; Kempster—unintentionally—broke it apart. Wingate had been successful in his relations with the common council; Kempster failed abysmally with the legislative body. Wingate left a congenial, achieving office; Kempster left chaos and confusion when he was impeached in the middle of his term.

[65] Wingate, "Smallpox in Wisconsin," pp. 271-272.
[66] For a statement about thinking on the issue of compulsory versus voluntary vaccination in the period, see C. S. Lindsley of the State Board of Health of Connecticut, quoted in Clark Bell, "Compulsory Vaccination; Should it be Enforced by Law?" *JAMA* 28 (1897): 49-53.

The new reform Republican mayor, John C. Koch, appointed Kempster in 1894 in an attempt to raise the health department above political intrigue. Other physicians, each backed by local supporters, had sought the position, but Koch chose Kempster because he did not seek the job, did not have a coterie loyal to him, and did favor civil-service reform. Kempster seemed well suited to the job of health commissioner even though he was new to Milwaukee and was best known for his work with the insane. He enjoyed a national reputation and had recently returned from investigating cholera abroad.[67]

The common council immediately questioned Kempster's appointment. Democrats might have been expected to oppose a Republican nominee, but Kempster's strongest opposition came from fellow Republicans, angry that their candidates did not get the job. Republican aldermen Robert Rudolph and Charles Kieckhefer, both from predominantly German wards, led the opposition. These men emphasized Kempster's lack of familiarity, because of his recent arrival in town, with Milwaukee's health problems. Kempster's English heritage and national stature acted against him in the minds of those who sought a local and possibly ethnic representative of Milwaukee's population. But the majority of the council decided to give Kempster

---

[67] Walter Kempster had only small ties to the City of Milwaukee. He had held posts at the State Lunatic Asylum at Utica, New York, and at the Northern Hospital for the Insane, at Oshkosh, Wisconsin. He had testified for the prosecution in the Guiteau case. He had moved to Milwaukee in 1890, but since that time had been on two missions abroad studying cholera and investigating Jewish emigration from Russia. His tenure in the City of Milwaukee, his critics were quick to point out, had been short indeed. For biographical material on Walter Kempster see *Dictionary of American Biography*, 324-325; Howard Kelly and Walter Burrage, *American Medical Biographies* (Baltimore: 1920), pp. 652-653; Watrous 2: 30-33; and most thoroughly, Sperry, pp. 56-69. See also Kempster's obituary, *Sentinel*, August 23, 1918; *New York Times*, August 23, 1918, p. 9. Kempster was buried in Arlington National Cemetery.

a chance, confirming his appointment by a vote of 23 to 13.[68]

With the nation and the city in the middle of a severe economic depression in 1894, the twenty-six patronage jobs controlled by the health commissioner assumed great significance. Although Kempster was a Republican, he was not susceptible to the influences of fellow party members. In fact, when he announced his appointees, it was clear that he had completely ignored the party's suggested lists of candidates. Council members immediately challenged the appointments and vetoed some of them. Their first encounter thus established the pattern of confrontation that characterized Kempster's relations with the common council throughout his years in office.[69]

As a result of this clash over department appointments, tension already existed between the health department and the common council in June 1894, when smallpox broke out in the city and an epidemic threatened. An atmosphere of cooperation, necessary to cope successfully with an emergency situation, was missing. Smallpox became the weapon with which certain members of the council, led by south-side saloon-keeper Robert Rudolph, fought the health commissioner. The Polish and German immigrant groups, hardest hit by the epidemic, furnished the movement's political strength.

Kempster reacted to smallpox in a manner similar to that of health commissioners before him. He at once hired extra physicians to launch a widespread vaccination campaign.

[68] *Sentinel*, April 17, 1894, p. 1; April 18, 1894, p. 2; April 24, 1894, p. 3; April 30, 1894, p. 4. Eight Republicans opposed and five Democrats supported Kempster. Without Democratic support, the Republican nominee would not have been confirmed. *Common Council Proceedings*, April 30, 1894. See also *Sentinel*, May 1, 1894, pp. 1, 4.

[69] Many of the applicants had excellent backgrounds, and had earned as much as $2,000 a year, but they were now out of work and were willing to drive garbage wagons. *Sentinel*, May 2, 1894, p. 3; *Sentinel*, May 29, 1894, p. 2; *Sentinel*, June 12, 1894, p. 3; June 26, 1894, p. 9; *Common Council Proceedings*, June 11, 1894, p. 132; May 14, 1894, p. 61.

He moved swiftly to isolate those patients reported to have the disease by removing them to the isolation hospital in the eleventh ward, acting under the 1892 ordinance giving him the power to do this forcibly if necessary. He enforced a strict quarantine on those allowed to remain at home. The department also carried on patient education campaigns and made wide use of the city's disinfecting van.[70]

The reaction of citizens to the health department's offensive was similar to the response during previous outbreaks. Most Milwaukeeans cooperated with the department, although the German and Polish areas of the city resisted vaccination and removal to the hospital. The health department had no authority to force vaccinations on anyone who did not want them, but it could prohibit nonvaccinated children from attending the public schools. Since the epidemic began just when the public schools were closing for the summer, this restraint was not used until school reopened in the fall.[71]

During June and July, smallpox appeared in all sections of the city, keeping the health department busy vaccinating and isolating reported cases. Kempster remained confident of the efficacy of his procedures and believed that the epidemic would not get out of hand. But by mid-July it was evident that a significant number of people were not cooperating with the health authorities. Many cases of smallpox went unreported. Discontent with health policy grew when smallpox seemed to localize in the south-side wards.

[70] MHD, *Annual Report*, 1894, pp. 26-32; MHD, *Annual Report*, 1895, pp. 10-12; MHD, *Annual Report*, 1894-1895, pp. 35-39; *Sentinel*, June 28, 1894; July 4, 1894, p. 4. On school closings, see the *Sentinel*, September 2, 3, 1894; November 3, 13, 24, 1894. See also *School Board Proceedings*, November 6, 1894, pp. 124-125.

[71] According to the *Sentinel*, at least two German-language newspapers argued that vaccination did not protect against smallpox and that its effects were often worse than the disease it was to prevent. The Anti-Vaccination Society continued to disseminate its information through pamphlets widely circulated in the city in three languages. *Sentinel*, August 1, 1894, p. 4.

Hardest hit by the contagion was the eleventh ward, site of the isolation hospital and home of Alderman Rudolph.

At first Kempster denied that there was a seat of infection in the south side, but the numerous cases discovered there belied his assertions. Significant numbers of parents refused to allow their children to be examined or vaccinated by health officials, which contributed to the rapid spread of the disease.[72] The immediate focus of the people's wrath was the isolation hospital in the eleventh ward. Despite the renovations that had recently transformed the institution into what health authorities insisted was a modern facility, residents in the neighborhood still viewed it as a pesthouse and the source of their trouble. They claimed it was a "menace to the health of citizens," a "slaughterhouse" where patients "were not treated like human beings." Health officials maintained that the hospital was in good condition and offered good service to the sick poor who were admitted.[73] Whatever the actual condition of the hospital in July and

[72] *Sentinel*, July 23, 1894. See also July 24, 1894, p. 3.

[73] *Sentinel*, August 3, 1894, p. 4. On July 23, Alderman Rudolph introduced an ordinance to remove the hospital and to purchase ground outside the city for the erection of a new one. Debate on this ordinance lasted throughout the summer and fall in the council. See *Common Council Proceedings*, July 23, 1894, p. 294. The city attorney was of the opinion that although the city could purchase land outside the city, and could build a hospital on it, the state law forbidding transportation of people with infectious disease would mean that no Milwaukeean could legally be taken to such a hospital. *Sentinel*, August 28, 1894, p. 3. The issue was finally dropped.

Dr. Kempster initially supported the removal of the hospital from the city because he saw in the move the possibilities of a great public facility. In August, with the present hospital full of smallpox patients, he realized the inadequacy of a thirty-bed hospital for an urban epidemic. He said: "The present hospital is an outrage, and the quarters are utterly inadequate for the proper handling of contagious diseases, and I am very glad that the Council contemplates the construction of a proper building for the purpose." *Sentinel*, August 3, 1894, p. 4. This position obviously added fuel to south-siders' claims, and he quickly altered it when he realized the use to which it would be put. He became a vocal advocate of leaving the hospital in its present location. See also *Sentinel*, August 4, 6, 1894; MHD,

August 1894, south-siders were convinced that it was a death house for those who went there as patients and that its presence infected the nearby districts.

Anxiety levels deepened over the summer, and on August 5 a crowd of neighbors successfully resisted an attempt by the health department to take a sick two-year-old to the isolation hospital. About 3,000 "furious" people armed with clubs, knives, and stones assembled in front of the child's house. Another of the family's children had recently died in the isolation hospital, and the mother was determined not to let the city "kill" another of her children. "I can give [her] better care and nourishment here than they can give [her] in the hospital," she claimed. "I will not allow my child to be taken to the hospital." Faced with the violent mob, and unable to control the situation, the ambulance beat a hasty retreat.[74]

There is no evidence that alderman Rudolph was in any way connected with the public outburst on the night of August 5. But by the next day his name was intimately linked with the south-side "rioters," and he was participating actively to organize and to mold the political force unleashed in his ward. On August 6 he introduced a resolution in the common council to remove the power of the health commissioner to take patients to the hospital against

---

Annual Report, 1894-1895, pp. 55-58. Most physicians took a position against removal. They did not think long-distance travel would be beneficial to an acutely ill patient, nor would it be convenient for physicians trying to deal with a city-wide epidemic. There also was disagreement about the condition of the present hospital, and the quality of care a patient could receive there. For Polish opposition to the hospital, see *Kuryer Polski*, July 23, 1894.

[74] *Sentinel*, August 6, 1894. See also *Evening Wisconsin*, August 6, 1894. For more on the August 5 episode, see MHD, *Annual Report*, 1895, pp. 42-43. For a popularized account of the riots see Richard L. Stefanik, "The Smallpox Riots of 1894," *Historical Messenger*, December 1970, pp. 123-128. Stefanik's Masters essay, "Public Health in Milwaukee: From Sanitation to Bacteriology," University of Wisconsin—Milwaukee, 1967, also deals with the episode.

Figure 12. "Smallpox Troubles in Milwaukee." *Leslie's Weekly Illustrated Newspaper,* September 27, 1894, *79:* 207. I am grateful to Martin S. Pernick and Janet S. Numbers for help in locating this illustration.

their will. Although the resolution had no effect against the ordinance that gave the power, the south-siders interpreted its adoption as a vote of support.[75] Rudolph appeared as a leader at public rallies, and on August 7, when a crowd gathered to protest not being allowed to attend the night burial of a smallpox victim, Rudolph addressed them with a speech that "was not entirely free from incendiarism." "I don't blame the people down here for being worked up," Rudolph told a newspaper reporter. "The patients at the hospital are not treated like human beings and the way the dead are buried is brutal."[76] Believing the rest of the city viewed them as "the scum of Milwaukee," south-siders insisted on their rights: "We will not submit to have our children dragged from our homes."[77]

[75] *Common Council Proceedings,* August 6, 1894, p. 326. See also *Sentinel,* August 7, 1894, p. 2, and MHD, *Annual Report,* 1894-1895, pp. 41-42.

[76] Quoted in the *Sentinel,* August 8, 1894, p. 1.

[77] Henry P. Fischer, eleventh-ward resident, quoted in the *Sentinel,* August 8, 1894. The *Daily News* tended to support the rioters, but not the

Kempster reacted by stiffening the official position, insisting that if smallpox spread it was not due to health department negligence but rather to the rioters themselves, who spread the contagion through south-side wards. He carefully defended every argument leveled at the department and told a reporter, "But for politics and bad beer, the matter would never have been heard of."[78] This deprecating comment did little to ease tensions between the health department and the south side. In fact, it emphasized the differences—class and ethnic—between Kempster and the immigrant south-side residents. Showing no compassion, Kempster remained firm during the entire episode, never bending to the south-siders, never recognizing that their concerns may have been legitimate. "I am here to enforce the laws," he said, "and I shall enforce them if I have to break heads to do it. The question of the inhumanity of the laws I have nothing to do with."[79]

The unrest on the south side clearly aided the spread of smallpox in that region of the city. Daily, crowds of people took to the streets, seeking out health officials to harass. Quarantine officials watching guard over houses were frequently the object of the mob's attack. With thousands of people roaming the streets and entering houses infected with smallpox, the contagion was destined to spread throughout the district. Case reports, despite the many concealed from authorities, indicated that south-side wards 11 and 14 were most severely hit during the summer and fall of 1894. (See Figure 13.)[80]

The mayor, finding the mass meetings on the south side "detrimental to the public order and safety and dangerous

riots in early August. "There is reason to believe that there has been some basis for the many criticisms . . . made." The *Sentinel* was quick to blame the "ignorance of the people" in not following existing health regulations and in defying authorities.

[78] Dr. Kempster is quoted in the *Sentinel*, August 7, 1894, pp. 1-2.

[79] Quoted in the *Daily News*, August 11, 1894.

[80] Out of a total of 1,074 cases, 846 (79 percent) were from south-side wards—half of those were from the eleventh ward.

to the public health," urged all citizens to "remain quietly in their homes."[81] Even Rudolph moderated his oratory after the chief of police warned him that his actions could be interpreted as inciting the rioters.[82] But still the mobs continued to resist health department attempts to stem the epidemic.

The focus of the crowds' hatred was Kempster, who symbolized arbitrary governmental authority that subverted immigrant culture and threatened personal liberty. Calling for his execution, crowds demanded that the "people's rights were paramount and should be protected, if need be, at the point of a pistol."[83] Women played a conspicuous part in the disturbances. Because police were reluctant to use their clubs on "feminine shoulders," women were effective at maintaining disorder. Armed with clubs and stones, they assaulted the city police sent to preserve order. They threw stones and scalding water at the ambulance horses in an effort to stop officials from forcibly removing any patients. As one newspaper observed the situation: "Mobs of Pomeranian and Polish women armed with baseball bats, potato mashers, clubs, bed slats, salt and pepper, and butcher knives, lay in wait all day for . . . the Isolation Hospital van."[84] (See Figure 12.)

[81] Quoted in the *Evening Wisconsin*, August 9, 1894, p. 1.

[82] The *Evening Wisconsin* put Rudolph's position this way: "Some of the more excited ones claimed that Rudolph had been bought up, but as a matter of fact there was another reason. It is understood that Alderman Rudolph was requested to appear yesterday afternoon at the office of the Chief of Police, where he met that official . . . and was told in plain language, it is said, that his words and actions had caused turbulence in the Eleventh Ward and that if the mass meeting which was to be held that night developed into a riot the blame would fall on him. He was given to understand . . . that even at that moment he was liable to arrest for the violent language he had used the night before" (August 9, 1894, p. 1). The English-language newspapers deplored the violence in the eleventh ward: "Law and Order Must Prevail," *Evening Wisconsin*, August 9, 1894; "The Law Must Be Respected," *Daily News*, August 9, 1894.

[83] *Sentinel*, August 9, 1894, p. 1.

[84] *Sentinel*, August 30, 1894, p. 3. See also August 10, 12; *Daily News*, August 10, 1894; MHD, *Annual Report*, 1895, p. 49.

City officials met daily to try to determine a course of action that would stem the riots and the spread of smallpox. They consulted the State Board of Health, whose secretary, Dr. U.O.B. Wingate, well understood Milwaukee's situation, and contemplated asking that body to take control of the city health department. But Wingate praised Kempster's efforts, indicating he would do nothing different, and left the city to its own authority.[85] Despite official efforts, the disturbances continued intermittently through August and into September.

Throughout the period of unrest, south-side citizen groups tried to reestablish the image of that section as a peaceable place, safe for business. Deeply regretting "the notoriety

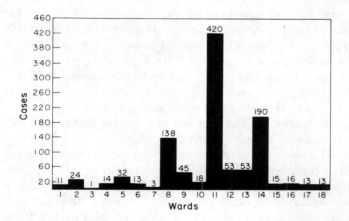

Figure 13. Milwaukee Smallpox Epidemic 1894-1895, Ward Distribution. Wards 5, 8, 11, 12, 14, and 17 were on the south side. *Source*: Milwaukee Health Department Annual Report, 1895.

[85] For details about the involvement of the State Board of Health, see *Wisconsin State Board of Health Report of 1893-1894*, "Report of the Executive Committee Relative to the Smallpox situation in the City of Milwaukee," pp. 56-69; Wingate, "Smallpox in Wisconsin"; and the MHD, *Annual Report*, 1894-1895, pp. 49-58.

which has been recently thrust upon that section of the city," one residents' meeting blamed the carelessness of the people for spreading the disease. Even the moderate groups, however, were not in sympathy with the health department, which, they felt, had lost the confidence of the people and was thus no longer competent to deal with health emergencies.[86]

The activity in the eleventh ward did diminish the effectiveness of the health department. Kempster himself admitted that, although smallpox had been under control before the riots, quarantine was impossible and the spread of the disease inevitable since the mobs began roaming the streets.[87] Although vaccinations were freely available, people refused them. During the violence on the south side the daily work of the health department virtually came to a halt. Patients could not be removed to the hospital in the face of weapon-wielding mobs; patients from other wards in the city could not be transported to the hospital through the hostile eleventh ward.[88] Citizens denounced Kempster for attempting to remove patients to the hospital, and further censured him when failure to remove such patients resulted in the spread of the epidemic.

To a meeting of concerned and sympathetic physicians and business people, Kempster vented his frustrations:

"The laws are not enforced because the Common Council has prevented me. Not a single proposition that I have made . . . has been acted upon. . . . Proposition after proposition has been made to revise the laws as they now are. This has caused opposition among the people. We come to a house to remove a patient and are resisted. They tell us that their alderman informed them that next week the laws will be changed and they need not go. I have been

---

[86] *Sentinel*, August 12, 1894; August 14, 1894.
[87] *Sentinel*, August 12, 1894, p. 1.
[88] *Evening Wisconsin*, August 29, 31, 1894; *Sentinel*, August 29, 1894, p. 3; *Sentinel*, September 1, 6, 1894, p. 1.

tied hand and foot with investigations, injunctions and work that is never finished."[89]

At the beginning of September, with the opening of school, the coming of cooler weather, and increased police action against the rioters, the roving mobs on the south side became less visible. The focus of the anti-Kempster movement changed from street action to the common council, as that body took up the battle in earnest.

Alderman Rudolph maintained a hold on his constituency and leadership of the mass movement against the health department by his actions and inflammatory rhetoric within the common council. There he introduced resolution after resolution and ordinance after ordinance, each one concerned with limiting the power of the health department and Walter Kempster. As a member of the council's health committee and a close friend of council president, William G. Rauschenberger, Rudolph had considerable influence over health measures.

Beginning with measures introduced on August 6 attempting to limit Kempster's power to remove patients to the hospital, Rudolph's actions crescendoed as the epidemic itself increased in intensity. In early September he introduced an ordinance designed to accomplish what his earlier resolution could not: legally to tie the hands of the health commissioner by not allowing him to remove patients without their consent. The ordinance passed the council and became law.[90]

[89] Quoted in the *Sentinel*, October 2, 1894, p. 3. See also October 3, 1894.

[90] *Common Council Proceedings*, September 4, 1894; October 1, 1894, November 26, 1894. See copy of the ordinance as passed in the Appendix to the 1894-1895 *Proceedings*, pp. 22-23. Section 1 provided: "But the commissioner of health shall not remove to any Isolation Hospital in said city any child or person suffering from any such disease who can be nursed and cared for during such illness in his or her home during the continuance of the disease except upon the recommendation and advice of the said commissioner of health or one of the assistant commissioners of health, and the physician, if any, attending upon such child or person,

Part of Rudolph's success in passing this ordinance is attributable to his scare tactics. The only member of the council in daily touch with the rioters, he promised renewed violence if the council did not pass his measure. He emoted loud and long on the injustice of tearing "a child from its mother's breast," convincing his fellow politicians that voters would not be happy unless this measure passed. Council president Rauschenberger, who agreed that Kempster was incapable of handling the epidemic, actively supported Rudolph.[91]

In October, with the epidemic still raging about the city, Rudolph called for a special investigating committee to inquire into Kempster's activities, listing thirty-four charges against him.[92] The main charges were that Kempster had been negligent of his duties in the management of the isolation hospital, that he showed ignorance of quarantine

---

not being a member of the health department of said city; and in case such commissioner, or assistant commissioner and such physician shall be unable to agree as to the advisability of removing such child or person, then they shall call in and appoint another physician not a member of the health department, and the decision shall be decisive of the question. The third physician called in, as above provided shall not receive or be entitled to any fees from the city for consultation or service in the decision of the case submitted to such board of physicians."

[91] *Sentinel*, September 5, 1894, p. 1. See also *Sentinel*, September 4, 1894; October 2, 1894, p. 9; October 6, 1894, p. 4; October 7, 1894, p. 4; December 2, 1894; *Evening Wisconsin*, October 8, 1894. See also MHD, *Annual Report*, 1894-1895, p. 69; *Common Council Proceedings*, September 4, 1894.

[92] *Common Council Proceedings*, October 15, 1894, pp. 524-531. Although the two actions discussed here—ordinance for removal of patients and impeachment—were not the only ones that Alderman Rudolph initiated against the health commissioner, they were the most important, and are therefore the focus of the discussion here. For additional measures that the council considered at Rudolph's initiation see *Common Council Proceedings*, August 20, 1894, pp. 364-365; September 4, 1894, p. 583. See also *Sentinel*, August 21, 31, 1894, p. 3; September 1, 1894, p. 5; October 2, 1894, p. 9; October 13, 1894, p. 5. See also *Evening Wisconsin*, August 30, 1894.

methods, that patients were removed from their homes when they could have been better taken care of at their residences, and that the health department had grown tyrannical.[93] Twenty-eight physicians, the majority of whom belonged to the regular local or state medical societies, signed testimonials of misconduct by the health commissioner. Twenty-four of them lived and practiced on the south side.[94]

Rauschenberger appointed Rudolph to chair the council's committee investigating the charges.[95] The impeachment proceedings were front-page news in all the city papers. Although most English-language papers declared that the investigation was a "farce" because it was led by so prejudiced a man, even the friendly *Sentinel* agreed that there was a need to clear the air.[96] The German-language

[93] *Sentinel*, October 16, 1894, p. 1.

[94] Of the 28, at least 19 were members of the local medical society, the State Medical Association, or the American Medical Association. See the membership list of the Milwaukee Medical Society and the local AMA members in the archives of the Milwaukee Academy of Medicine. The membership list of the State Medical Society was in the *Wisconsin Medical Journal* vol. II (1903-1904), pp. 196-208. At least two of the physicians had been brought to court by Kempster and charged with not reporting smallpox, a fact which may account for their hostile position. For location of the physicians, see the *Milwaukee City Directory*, 1894. I would like to thank William J. Orr, Jr., for locating the addresses and wards of the physicians. One of the physicians cannot be located.

[95] The rest of the investigation committee were Aldermen Derwerth (ninth ward), Muenzberg (fifteenth ward)—later replaced by McGarigle (first ward)—Lindsay (fifth ward), and Weiher (Democrat from the eighteenth ward). Aldermen Lindsay and Muenzberg were not eager to serve. Muenzberg was quoted saying: "They can carry me to the courthouse in a patrol wagon or impose fines upon me, and still I will not act. I will have nothing to do with the investigation." *Sentinel*, October 23, 1894, p. 9. Muenzberg did get himself off the committee, but ultimately voted for impeachment. See also *Sentinel*, October 19, 27, 1894, and *Daily News*, October 16, 1894.

[96] See the *Sentinel*, *Daily News*, and *Evening Wisconsin*, passim, October 15, 1894 to February 21, 1895. See, for example, *Sentinel*, editorial of January 5, 1895, entitled, "The Farce Continued." Other newspapers voiced similar sentiments. The *Catholic Citizen* ridiculed the manner of

press, on the other hand, endorsed Rudolph and fully supported the impeachment proceedings for the "autocratic,
obstinate, freakish and brutal officer."[97] The one exception
to the German-language press support of Rudolph came
from the Socialist *Vorwärts*, in which editor Victor Berger
condemned both Kempster and Rudolph. Berger voiced
greatest scorn for "the saloon-smallpox Alderman" Rudolph, who was taking advantage of "the stupidity and ignorance of certain elements on the south side" for political
advancement. The *Vorwärts* supported Kempster's policies
of isolation and vaccination but condemned the health commissioner's insensitive handling of the epidemic.[98]

Although no ethnic newspaper condoned violence *per se*,
many did understand the south-siders' desire to "reply to
violence with violence." The editors echoed the sentiments
of the south-siders, who felt that Kempster, motivated by
excessive legalism, had not treated the sick with compassion
and had allowed his agents to commit physical harm to
person and to property in order to follow the letter of the
law. The Polish press remained more moderate than the
German, but it also did not support Kempster.[99]

Rudolph's allies in his anti-Kempster activity, then, included the ethnic press, the anti-vaccinationists, and the
south-side activists. There were many physicians among his
supporters, including Ernst Kramer, who had earlier opposed Johnson's vaccination campaign, A. B. Farnham, an

---

putting the "prosecuting attorney on the bench as the judge of the case."
October 20, 1894, editorial. Similar sentiments were voiced by other newspapers. See also, *Sentinel*, October 24, 1894, p. 4.

[97] The *Seebote* quoted in the *Sentinel*, September 2, 1894, p. 4.

[98] *Wisconsin Vorwärts*, August 7, October 15, November 27, December
7, 1894. I am grateful to William J. Orr, Jr., who searched this and other
German and Polish newspapers' reactions to Kempster and translated
them for me.

[99] *Germania*, August 17, 1894; *Abendpost*, August 10, 1894; *Herold*, August 9, November 2, 1894; *Kuryer Polski*, August 9, 11, 13, 14, 1894,
September 1, October 19, 31, 1894.

active Milwaukee Medical Society member, and R. E. Martin, son of the former health commissioner.[100] Their leader was Emil Wahl, a German physician with a successful southside practice. Wahl was a member of the Milwaukee Medical Society (although he resigned from that organization when it accepted Kempster as a member), and his accusations that Kempster was incompetent to deal with the epidemic carried weight in the community.[101]

The epidemic reached its height during the month of October, when the impeachment hearings began. Testimony revealed some of the issues dividing Milwaukee's physicians.[102] Just as physicians had differed among themselves earlier on the value of vaccination, they came to

[100] See the signers of the petition to remove Kempster from office, *Sentinel*, October 16, 1894. For interviews and testimony from these physicians against Kempster, see *Abendpost*, August 29, October 24, 1894; *Herold*, August 13, October 19, 1894; *Kuryer Polski*, October 19, 1894. A. B. Farnham, speaking at a citizens' meeting, decried Kempster's "lack of tact and sympathy, his want of knowledge and prescience—in short, his mismanagement of the office," concluding that it would be impossible for Kempster to "regain the confidence of the community." Quoted in the *Sentinel*, October 25, 1894.

[101] For biographical material on Emil Wahl, see Frank, *Medical History*, p. 72. He appears on the membership list of the Milwaukee Medical Society, having joined the organization in 1888. However, the society was clearly in the Kempster camp on the investigation. Rudolph's name was hissed at a meeting held on October 22 between the medical society and the business leaders of the city, and Kempster's loudly cheered. The joint meeting unanimously passed a resolution to the effect that it was not deemed advisable to change any of the existing health laws, and appointed a committee to attend the impeachment proceedings. *Sentinel*, October 23, 1894, p. 3. Dr. Wahl resigned from the Milwaukee Medical Society when it admitted Walter Kempster to membership. Minutes, January 22, 1895, February 12, 1895.

[102] The transcripts of the impeachment hearings appear to be lost, and it is impossible to discern all the technical aspects of the medical debate from the newspaper accounts. The hearings can be followed in any Milwaukee newspaper. See especially the *Sentinel*, October 28, November 15, 16, 17, 28, 29, December 4, 5, 1894 and January 4, 5, 6, 11, 12, 15, February 14, 15, 16, 19, 1895; and the *Journal*, January 4-10, 1895.

111

opposing conclusions about defining proper smallpox treatment. Emil Wahl may or may not have been politically motivated to speak publicly against Kempster, but, either way, his medical disagreements were real. Wahl's principal complaint was that Kempster discharged patients from the isolation hospital while they were still contagious, thus aiding the spread of the disease. The debate centered on the contagiousness of the smallpox pustule at its various stages. A number of physicians charged that Kempster had released patients whose pustules still oozed, allowing them to spread the disease throughout the city. The testimony revealed that practicing physicians disagreed and were confused about how to interpret the stages of smallpox. Physicians argued particularly about the value of the disinfecting baths used at the isolation hospital and whether or not they could prevent the spread of infection.[103]

Kempster cross-examined many of the witnesses, attempting to show that they, not himself, were unfamiliar with smallpox and unable to recognize its contagious states. He continued to deny that he had acted wrongly, maintaining that isolation, disinfection, and vaccination, the accepted responses to smallpox, were the basis of his policy. The State Board of Health and the Milwaukee Medical Society supported Kempster, and their support obviously helped the health commissioner to withstand the public condemnation.[104] But, however acceptable the public found Kempster's stand theoretically, his insensitivity and inflexible behavior as revealed in the daily testimony became harder and harder to defend.

Rudolph's supporters testified to seemingly endless cases

---

[103] Kempster claimed that the disinfecting baths given to patients when they were discharged from the hospital prevented them from spreading smallpox. Farnham thought that the antiseptic baths were "a criminal innovation." *Sentinel*, November 16, 1894. Wahl believed in antiseptics "to a limited extent." *Sentinel*, November 17, 1894. See also November 15, 28, 29, December 4, 5, 1894.

[104] *Sentinel*, October 23, 1894.

illustrating the neglect, mismanagement, and "brutality" of the public health officials.[105] South-siders attested to the drunken, malicious behavior of quarantine officers, disinfecting crews, and ambulance personnel. Mothers told of how their children had been forcibly wrested from their homes, some when they were still wet from their baths or inadequately clothed for travel. Witnesses claimed that some smallpox victims who lived in crowded tenements had gone untreated by the health department, whereas others who could best be isolated at home were cruelly ensconced in the hospital. One ex-patient complained that he had been removed to the hospital against his will when he did not even have smallpox. Hospital attendants and former inmates testified that Kempster himself never visited the hospital, in which patients received inadequate and sometimes dangerous treatment. They said that attendants administered medicine to all from a single spoon, that the unscreened windows allowed flies to carry the disease into the neighborhood, and that the nursing staff was too small to care for the numbers admitted each day. Kempster's refusals to hear any complaints or to sympathize with the hardships that his policies created were catalogued over and over again.[106]

When they had heard all the testimony, the investigating aldermen judged that nine of the original thirty-four charges had been sustained, and they recommended conviction. The most compelling charge was that Kempster had provoked "riot and disturbance by his absurd pretensions and arbitrary and unjust methods."[107] Speculation ran high in the city about the council vote on the impeachment questions, with newspapers predicting the decision. Opinion shifts of some aldermen made front-page headlines. The

---

[105] *Germania*, August 17, 1894; *Herold*, October 23, 1894.

[106] See especially *Sentinel, Abendpost, Journal*, and *Kuryer Polski*, November 1894 to February 1895.

[107] *Sentinel*, February 22, 1895.

council heard the testimony—selected for their ears by Rudolph himself—for three days and nights consecutively. The exhausted aldermen were eager to be finished with the whole business. As the council neared the vote, its president complained: "I am at a loss to know whether we are attending a circus or a session of the Common Council." Excitement ran high, interruptions punctuated the proceedings, and disorder prevailed. Finally, in February 1895, with the epidemic not yet over, the council voted 22 to 14 to dismiss the health commissioner.[108]

The *Sentinel*, while admitting that Kempster's "usefulness" had been diminished, expressed outrage at the conclusion of proceedings that had been "disgraceful in the extreme on account of their malicious and grotesquely unjust character."[109] Other English-language newspapers echoed these sentiments. "Milwaukee may well bow her head in shame over the vote in the council on Dr. Kempster," editorialized the Sunday *Telegraph*.[110] But the German papers rejoiced at the outcome. Regretting that the action had not been accomplished sooner, the *Abendpost* voiced its pleasure that seventeen German aldermen had voted in favor of impeachment while only two had supported Kempster.[111] The *Herold* also greeted the decision with "satisfaction."[112]

Although the impeachment vote did not divide along party lines, the move against Kempster was nonetheless political. Patronage, class, and ethnic divisions created a significant part of the opposition. Physicians who sought

[108] For the press assessment of how the vote would be, see the *Sentinel*, February 14, 15, 16, 18, 1895; *Journal*, February 18, 1895; and the *Daily News*, February 18, 1895. See the description of the final session in the *Journal*, February 21, 1895 and the *Sentinel*, February 21, 1895.

[109] *Sentinel*, February 22, 1895.

[110] *Telegraph*, February 23, 1895. See also the *Daily News*, February 22, 1895; *Evening Wisconsin*, February 22, 1895.

[111] *Abendpost*, February 22, 1895.

[112] *Herold*, February 22, 1895. I am grateful to Edith Hoshino Altbach for translating this editorial for me.

his post or appointments within the health department were disappointed and resentful of the man who held office. Kempster had particularly alienated the south-side physicians with his cavalier attitudes toward that region's suffering and with his exclusion of south-siders from most health department jobs. Republican politicians were bitter because of their loss of influence in department appointments.[113] From the beginning of his tenure in office, Kempster had labored against a vocal opposition. When the epidemic struck the city, the opposition used it as a weapon successfully to resist the health officer. The issue of the isolation hospital carried with it a tradition of anxiety and fear and was easily employed against Kempster. The issue of forcible removal of children from their homes pulled at the heartstrings of every immigrant parent. Kempster's position in Milwaukee, whatever its medical merits, was politically untenable.

The ethnic divisions in the city were evident from the press differences as well as from the impeachment vote. Aldermen from those wards which contained large numbers of German and Polish immigrants, or their descendants, sustained the impeachment vote. Those whose wards were largely populated by American-born or by immigrant groups other than German and Polish voted for Kempster. Party affiliations did not determine the vote, as both Democrats and Republicans on the council split. Figure 14 illustrates the close correlation between German and Polish ethnicity and the vote against Kempster.[114]

---

[113] Walter Kempster adopted this explanation. *Sentinel*, October 16, 1895, p. 1. For other contemporary arguments along these lines, see *Sentinel*, August 30, 1894, p. 3; August 31, 1894, p. 4; September 2, 1894, p. 4; September 4, 1894, p. 4; February 20, 1895, p. 3; February 22, 1895, p. 4; *Daily News*, August 31, 1894; *Journal*, February 19, 1895; *Catholic Citizen*, September 8, 1894, p. 4; *Sentinel*, January 21, 1896.

[114] The sources for ethnicity were the *Eleventh United States Census*, 1890 and the 1895 Wisconsin State Census. Precise population figures by ethnic group were not available by ward.

The smallpox epidemic of 1894 and 1895 severely divided Milwaukee, creating long-lived political conflicts and resentments. It had a retrogressive effect on public health in the city. Traditionally, epidemics led to an increase of health department power. But during this epidemic the opposite happened. During the height of the epidemic, with fear running high, the common council repealed measures that health officers thought effective in halting the spread of the disease and fired the physician who supported those techniques. As Figure 15 illustrates, the moves against Kempster were all initiated while the epidemic raged in the city.

Kempster did not willingly leave office. He remained at

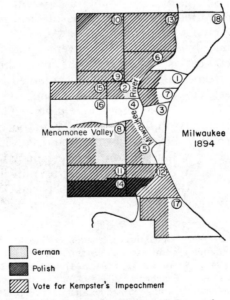

Figure 14. Ethnic vote for Kempster's Impeachment. *Sources:* Milwaukee Health Department Annual Reports and Proceedings of the Milwaukee Common Council.

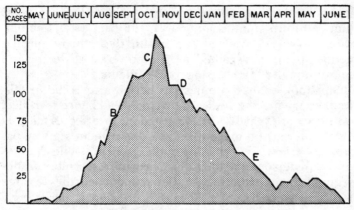

REPORTED CASES

Figure 15. Milwaukee Smallpox Epidemic, 1894-1895, reported cases and moves against Kempster.
A. August 6, 1894. Resolution on forcible removal.
B. September 4, 1894. Ordinance on forcible removal.
C. October 15, 1894. Resolution for impeachment investigation; revised ordinance on forcible removal.
D. November 26, 1894. Revised ordinance on forcible removal passed.
E. February 21, 1895. Common Council impeaches Health Commissioner.
Sources: Milwaukee Health Department Annual Reports and Proceedings of the Milwaukee Common Council.

his desk, carrying out the functions of the health commissioner for a week after the impeachment vote, until he was forcibly ejected from that office by two "big" police officers on February 28, 1895. Dr. H. C. Bradley, his assistant commissioner, took over the position.[115] Kempster appealed the

[115] See accounts of the leave-taking and Kempster's protest speech in the *Sentinel*, March 1, 1895; *Daily News*, February 28, 1895. For Kempster's reaction to the impeachment proceedings, see MHD, *Annual Report*, 1894-1895, pp. 76-80. There is some evidence to suggest that Bradley had been acting as health commissioner before the impeachment, indicating perhaps that Kempster was not fully functioning during the disturbance. For example, Kempster usually communicated with the schoolboard about

117

legislative decision, and after one year the courts reinstated him as health commissioner with full back pay. The episode, however, affected the health department permanently. It never regained all the powers lost during the tumultuous 1894-1895 smallpox epidemic.[116]

Smallpox entered a quiescent period after 1895 in Milwaukee, giving the health department a chance to reflect on its actions. Civil-service legislation, in effect after July 1895, altered the makeup of the department staff as patronage gave way to training and experience. The changes in health department operations became discernible in subsequent smallpox outbreaks. Department activities became less controversial as public health professionals increasingly removed themselves from political influence.

Through mild outbreaks of smallpox in the first decades of the twentieth century, health officials continued to advocate vaccination and isolation, and support for their programs gradually grew. Whereas Health Commissioner Schulz had felt "absolutely powerless to properly contend with the disease" in 1899,[117] Commissioner Bading in 1909 "en-

school closings, but on December 20 Bradley took over that function. *Proceedings* of the Milwaukee School Board, November 6, December 20, 1894.

[116] *Sentinel*, January 28, 1896, p. 1; see also *Daily News*, January 27, 1896. MHD, *Annual Report*, 1894-1895, p. 86. For the salary debate, see the *Sentinel*, January 31, 1896, p. 3; February 4, 1896, p. 2; March 3, 1896, p. 7; see also the *Common Council Proceedings*, January 30, 1896, February 3, 1896, March 2, 1896, August 31, 1896, p. 447, April 11, 1898, p. 876; MHD, *Annual Report*, 1894-1895, p. 87; MHD, *Annual Report*, 1896, p. 55. An additional blow to the health department came from the Wisconsin Supreme Court in 1897. In *Adams v. Burdge* the court ruled that the State Board of Health had no authority to require vaccination of children as a condition of attendance at school in the absence of state legislation to that effect. Seventy N. W. 347; 95 Wis 390; 37 L.R.A. 157; 60 Am. St. Rep. 123. See William Fowler, *Smallpox Vaccination Laws, Regulations and Court Decisions* USPHS, Supplement No. 60 to the Public Health Reports (Washington: U.S. Government Printing Office, 1927), p. 73; and James A. Tobey, *Public Health Law*, 3d ed. (New York: The Commonwealth Fund, 1947), pp. 235-250.

[117] MHD, *Annual Report*, 1899, p. 9. Letter from F. M. Schulz, Health

countered less opposition to [vaccination] than in former years."[118] When the Socialists won the municipal elections in 1910 and fulfilled their promise to build a new city hospital, the health department felt even more secure. Thus, by the time Health Commissioner John P. Koehler planned his attack on a virulent smallpox outbreak in 1925, the city had regained some of the powers lost in 1895 and was ready for him to show what the department had learned since the fateful outbreak thirty years earlier.

Koehler's attack on smallpox incorporated the same principles that had governed Johnson and Kempster before him: vaccination and isolation. But Koehler, unlike his predecessors, realized that "people cannot be vaccinated against their will," and he thus emphasized public education. In March and April 1925 Koehler vaccinated 62,000 citizens; in May, when he stepped up the pressure tactics in the media and through the business community, 233,000 Milwaukeeans came forward for inoculations.[119]

The education campaign employed warnings in the newspapers, posters on the streets and in stores, bulletins in factories, and letters to all major firms. The health department also used radio spots to aid the cause of public health. When increasingly adverse conditions warranted stronger measures, Koehler incorporated the fear element into his education work. Posters showing cases of virulent smallpox appeared all over the city, movie houses issued warnings on the screen, and the health department issued daily bulletins vividly advertising existing cases and deaths. Under a threat of quarantine Koehler pressured employers to vaccinate their workers. He quarantined six blocks of the city in those areas of the south and west sides most heavily infected, not allowing residents to leave their houses

---

Commissioner to the Common Council, *Common Council Proceedings*, January 3, 1899, p. 614.

[118] MHD, *Annual Report*, 1909, p. 13.

[119] John P. Koehler, "How Milwaukee aborted its smallpox epidemic," *Wisconsin Medical Journal* 24 (1925): 324.

Figure 16. School children line up for smallpox vaccination. Courtesy of City of Milwaukee Health Department.

unless they had been vaccinated. He enlisted the help of churches, clubs, settlement houses, and other community-based groups, and boasted a record vaccination total of 423,000 people by the time the epidemic receded. He achieved "almost universal cooperation," Koehler realized, because "of many conferences with representatives of various interests."[120]

Medical policy about how to limit the adverse effects of smallpox had not changed significantly between Johnson and Koehler, despite widespread acceptance of the germ theory in the intervening period. Vaccination and isolation remained the mainstays of prevention. But the health department had changed its reactions to smallpox in the years since it first encountered the disease. Understanding how

[120] *Ibid.*, p. 326. See also *Sentinel* and *Journal* March-June 1925, and the *Bulletin of the Milwaukee Health Department* November 1924 and March-June 1925.

best to implement traditional policies and how to marshall community forces for effective health controls had undergone a metamorphosis, particularly between the 1890s and the 1920s. Kempster had feared the forces that opposed him and the people whose lives and language differed from his own and had tried to suppress the differences through force and intimidation. Koehler, realizing that change could come only through mutual understanding, worked with the various groups in the city to encourage cooperation rather than resistance. He was not above using pressure tactics, such as threatening business with shut downs, but he always increased his demands gradually and worked with professional and community groups to insure support. Koehler benefited, too, from the added help of a staff selected through merit examinations rather than through political tests and an epidemic that struck during a time of economic prosperity rather than of depression.

Fighting smallpox had taught Milwaukee public health officials a lot about controlling public health problems. The lessons learned, while relatively insignificant in lowering the city's overall mortality, helped to establish health department authority in the city, gave visibility to department activity, and influenced other aspects of public health work. Sanitation campaigns, as we will see in the next chapter, found infectious diseases to be useful allies.

CHAPTER FOUR

The Politics of Health Reform: Garbage

One winter morning in 1892 Milwaukee resident James Holton stumbled out of bed and turned on his water faucet. Out came "very dark-colored" material filled with "a large number of brown and green globules of gelatine-like substance." Waking up faster than he might have otherwise, Holton broadcast his finding to his family, his neighbors, and—on the front pages of the daily press—to municipal authorities.[1] Other Milwaukeeans experienced similar shocks in the ensuing weeks and bitterly complained that the garbage dumped in Lake Michigan ruined the quality of the city water. Garbage-contaminated water became a major issue in the spring aldermanic elections, and throughout the decade the questions of why the city was unable to provide a potable liquid or to dispose of its garbage in a sanitary manner reverberated through the halls of the health department, the common council, and the municipal courts.

Garbage repeatedly polluted Milwaukee's water supply; it also littered the city streets, mixing with horse manure and other street droppings to create an irritation and of-

[1] Milwaukee *Sentinel*, January 27, 1892, p. 1. See also January 28, 30; February 2, 28, 1892.

122

fense to all who tried to maneuver in the crowded thoroughfares. In the summers, piles of rotting wastes sizzled under the hot sun, reminding everyone of the negative aspects of urban life. Not only was the filth unpleasant, but many urbanites thought it was responsible for the high amount of sickness in the city. The popular miasmatic theory of disease taught that foul atmospheric conditions and excessive filth caused disease. Physicians hoped that cleaning the streets and picking up the garbage would improve the health of city residents.[2] With mortality rates high (over 26 per 1,000 in some years) the city could not afford to ignore the sights that it witnessed every day. "Take a stroll of a few blocks in almost any part of Milwaukee," urged an observer in 1892, and you will find "heaps of dirt . . . broken pavements, breakneck crossings and uninviting pools of filth."[3] Milwaukee's streets, those "perfect avenues of swill," became the focus of much citizen attention; cleaning them and making them both passable for traffic and safe for health became one of the biggest challenges that municipal government faced—and ultimately conquered—in the Progressive period.[4]

Ashes, manure, garbage, night soil, sewage, rubbish, and dead animals littered urban streets, but garbage alone caused the biggest dilemma for officials trying to improve Milwaukee's health statistics. Responsibility for garbage rested with health officials, who frequently found, as one of them said, that no other health problem gave "more trouble and annoyance than this question, that like a nightmare haunts

[2] Even after the general acceptance of germ theory in the 1890s, ridding the city of gross filth continued to be the focus of public health officials. For a discussion of the popular miasmatic theory, see George Rosen, *A History of Public Health* (N.Y.: MD Publications, 1958), pp. 287-290; and Richard H. Shryock, *The Development of Modern Medicine* (N.Y.: Hafner Publishing Co., 1969), pp. 84f, 219f.

[3] *Sentinel*, May 30, 1892, p. 4.

[4] The quote is from the *Sentinel*, January 22, 1869, p. 1, but similar descriptions abound into the twentieth century.

the health officer."[5] Before 1911, when physicians, having solved the acute problems, surrendered the garbage problem to the engineers, Milwaukee had tried all known methods of disposal: feeding it to swine, filling land, fertilizing farms, rendering, cremating, and dumping it in the lake. All the methods caused problems, tried the patience of the public health physicians, and sorely tested the ability of the municipal legislature to cope with the sanitary problems of the quickly growing city.[6]

Before 1875 individual householders disposed of their own garbage. Usually they left their wastes for the hogs that roamed the streets or the "swill children"—commonly immigrant youngsters trying to supplement the family income—who collected the most desirable kitchen refuse that Milwaukeeans produced.[7] Obviously unequal to the task of collecting the wastes of an entire city, these "little garbage

[5] Robert Martin, "Disposal of Garbage at Milwaukee," *Public Health: Papers and Reports of the APHA* 15 (1889): 64.

[6] Rudolph Hering, the national sanitation expert, commented after studying Milwaukee's garbage situation that the city had "the most varied experiences in the disposal of its garbage of any city in this country." "Report on Garbage Disposal," submitted to the Milwaukee Common Council, December 23, 1907. Reprinted in the MHD, *Annual Report*, 1907, pp. 140-188, quote from p. 141. The manure problem in cities was not inconsiderable. In Chicago, for example, it was estimated that 100,000 horses averaged 750 tons of manure every 24 hours—this figure did not include the stockyards. See "Report of the Committee on Disposal of Waste and Garbage," *Public Health: Papers and Reports* 17 (1891): 97. It was estimated that one average human produced 57 pounds of fecal excreta and 114 gallons of urine annually. Alfred Ludlow Carroll, "Disposal of House Refuse," *JAMA* 14 (1890): 231. For nineteenth-century definitions of these terms see, for example, J. Berrien Lindsley, "On the Cremation of Garbage," *JAMA* 11 (1888): 514; "Report of Committee on Disposal of Waste and Garbage," *Public Health: Papers and Reports* 17 (1891): 95; "Report of the Committee on the Disposal of Garbage and Refuse," *ibid.*, 23 (1897): 207-208.

[7] For a brief account of swill gatherers in Milwaukee, see Charles D. Goff, "The Swill Children of Milwaukee," *Historical Messenger* (March 1960), pp. 9-11. See also MHD, *Annual Report*, 1873, pp. 17-18; *Sentinel*, June 16, 1873 and July 13, 1878.

124

gatherers" left the "backyards and alleys . . . reeking with filth, smelling to heaven."[8] Alarmed by the high urban death rates that he thought resulted from such unsanitary conditions, Milwaukee's first health officer, Dr. James Johnson, waged a campaign to convince the aldermen to take over the garbage collection from the children. Johnson received some support in the press, where one letter writer, realizing that "the first duty of every city is to guard the health of its people," urged the city fathers to "accept the responsibility that devolves upon [a growing city] and institute systematic collection."[9] The aldermen remained reluctant to spend precious city funds on so mundane a task, but finally allowed themselves to be persuaded on condition that the choice of whether or not to contract for garbage removal rest in the individual wards.[10]

Five (out of thirteen) of the most populous wards applied for ward collection in 1875.[11] Only three sought the service the following year, and by 1878 none applied for the garbage contract.[12] The first experiment in urban responsibility proved a failure. Because of old loyalties to the swill gatherers, many residents refused to give their refuse to the city collector, maintaining that they could "give it to

[8] Health officer Dr. James Johnson referred to the children this way in MHD, *Annual Report*, 1875, p. 46. The street description is quoted from a letter to the *Sentinel*, July 10, 1873, p. 8, signed "Pro Bono Publico."

[9] *Sentinel*, June 16, 1873. See also MHD, *Annual Report*, 1873, pp. 17-18.

[10] MHD, *Annual Report*, 1873, p. 7; *Sentinel*, July 10, 1873, p. 8; July 13, 1873, p. 8; July 31, 1873, p. 8; August 9, 10, 1873, p. 8; May 26, 1874, p. 8; MHD, *Annual Report*, 1873, pp. 47-48; *Proceedings of the Common Council*, May 4, 1874, p. 14; May 21, 1874, p. 35; May 25, 1874, p. 37.

[11] *Sentinel*, December 23, 1874, p. 8; March 13, 1875, p. 1; March 17, 1875, p. 8; May 26, 1875, p. 8; *Proceedings of the Common Council*, February 2, 1875, p. 229; MHD, *Annual Report*, 1875, p. 45; *Sentinel*, March 11, 1876.

[12] *Sentinel*, July 6, 1878, p. 8. The wards that applied for collection in 1875 were one, two, three, four, and seven, the most populous in the city. In 1876, wards one, four, and seven contracted with garbage collectors.

whom they please."[13] Thus compliance varied and the collector never knew whether or not he would find garbage in a particular neighborhood. If unable to obtain full loads, he could not sell them profitably to the farmers. Despite the ward contracts, "most of the garbage . . . is removed by boys and girls and women, mostly of Polish nationality, who use the material collected to feed hogs."[14] Aldermen ceased supporting the collection service when they realized that the voters did not appreciate it. By the end of Johnson's tenure as a city health officer, there was no city garbage service.

During the summer of 1878, as the sun beat down on almost 100,000 people in Milwaukee, the new health commissioner, Orlando Wight, found garbage "accumulating all over the city, creating an offensive unwholesome nuisance. The stuff is thrown in the alleys or stands, full of maggots and stinking, in swill barrels."[15] The situation cried out for a more effective municipal response.

The children who spent their days and nights collecting garbage particularly offended Wight's sensibilities. "The poor little fellows," he declared, "ought to be at school learning to become good American citizens."[16] Wight felt a certain amount of sympathy for "a few of the poor" who would suffer if the city replaced swill children with an efficient city collection system, but he nonetheless pressured council members to take complete responsibility for city garbage. As a result of the campaign that Wight initiated in the health department, the council, deciding that the initial problem had been the ward basis of the previous system, voted to let one city-wide contract for removing garbage and to restrict the activity of hogs and children.[17]

[13] *Sentinel*, July 16, 1875, p. 1. See also July 14, 1875; MHD, *Annual Report*, 1875, p. 46, and 1876, p. 66.

[14] MHD, *Annual Report*, 1877, p. 24.

[15] Dr. Wight is quoted in the *Sentinel*, July 13, 1878, p. 3.

[16] *Ibid.*

[17] MHD, *Annual Report*, 1878, p. 238. The two ordinances were reproduced on pages 29-33, 57-58. See also the *Sentinel*, July 20, 1878, p. 5;

The enforcement of these actions did hurt the poorer immigrant families who relied heavily on the swill trade. These working people resented the subsequent decrease in their income, and, in the words of their daily newspaper, complained "it is a great pity if [our] stomachs must suffer to save the noses of the rich."[18] The potential health benefits of the measures that took away an important source of income remained elusive to them.

But the swill gatherers did not have enough political strength to stop an increasingly active health office, which put the informal system out of business and awarded the first city-wide garbage contract in 1878.[19] The yearly contracts for collection, monitored by the city physician, continued until 1886. The health officer fought the battle for cleanliness assiduously, but he did not have an easy task. In 1879, for example, a confusion over the bidding process delayed the signing of a garbage contract, and for six weeks of unusually warm autumn weather no one picked up Milwaukee's garbage. When a contractor finally resumed the job, a "swell of indignation" arose among citizens who claimed that the company did not adequately perform the job. The distraught citizens "demanded" relief from the responsible health officials.[20] They complained that either collectors did not pick up the garbage regularly or that they dumped it indiscriminately on vacant lots. Collectors, who wanted only fresh refuse to sell to farmers for pig food or fertilizer, frequently refused to collect the messiest of the city's wastes,

---

July 12, 17, 20, 27, 1878; August 27, 1878, p. 3; September 3, 1878, p. 2; September 5, 1878, p. 3; September 6, 1878, p. 4. See also Goff, who noted that the ordinance provision "it shall be unlawful for any person to interfere in any manner with the collection and removal of garbage and offal" was specifically aimed at the swill children, although they were not named.

[18] *Daily News*, June 11, 1879.

[19] For statistics on numbers of loads collected from each ward see MHD, *Annual Report*, 1878, pp. 331, 334; and MHD, *Annual Report*, 1879, pp. 261, 301.

[20] *Sentinel*, December 9, 1879, pp. 4-5.

especially when citizens did not separate ashes from garbage.[21] Wight realized that whether the problem originated with the collector or the householders, Milwaukee remained garbage-littered and dirty. "We are doing the very best we can," he lamented, "with a very limited force."[22] The frustrated public physician, dealing with twenty-five to thirty garbage complaints daily, longed for a more efficient service that could relieve Milwaukee of its unsanitary burdens and make the city a healthier place in which to live.

In 1886 the garbage situation reached crisis proportions. Citizens of the outlying districts in which the collectors had disposed of the city garbage, fearing the "poisoned atmosphere" arising from the garbage-laden ground and the unwholesomeness of garbage-fed pigs, summarily stopped allowing the city dump carts on their property. City collectors thus had no place to put their loads, and the "great question" became what to do with the increasing mounds of waste produced in Milwaukee every day.[23]

Two enterprising Milwaukeeans developed a "solution of this vexed garbage question" and convinced Health Commissioner Robert Martin, Wight's successor, of the health value of cremation, or total destruction by fire. William Forrestal, a veteran contractor, and Frederick A. Marden, a physician interested in sanitation problems, formed the Phoenix Garbage Cremator Company and, as the legal difficulties of disposing of city garbage mounted, won the city

[21] Health Commissioner Wight attacked the citizens who did not separate their garbage in a letter to the editor of the *Sentinel*, December 10, 1879, p. 2. The defensive tone of the letter indicates one reason why Dr. Wight had less than the full support of the city behind him.

[22] *Sentinel*, December 13, 1879, p. 3. For more on this particular crisis, see the *Sentinel*, October 8, 1879, p. 4; October 9, 1879, p. 8; October 30, 1879, p. 8; November 10, 1879, p. 8; and the MHD, *Annual Report*, 1879, pp. 217-218.

[23] *Sentinel*, November 4, 1886. See also editorial, June 8, 1885, p. 4; and September 16, 1886, p. 4.

contract in 1887.[24] For the first time the city paid for both collection and disposal of all city garbage. (See Table 4-1.)

Until the opening of the disposal plant, Commissioner Martin proceeded "with neatness and dispatch" to organize garbage collection within the health department, initially trying to bury it on farms outside the city. But the surrounding towns refused even temporary permission and forced Martin to dump the offal into Lake Michigan, the source of city water.[25] During the few months when Martin emptied the garbage into the lake, residents protested that it was "positively unbearable," a health hazard, and a potential threat to the city water and fish supply.[26]

Citizen protests grew even louder when the new crematory opened in September 1887. One "stroller" through the city noted that it produced "a killing odor . . . with strength enough to paralyze the man in the moon."[27] Even more discouraging, however, was the fact that the new furnace did not efficiently destroy its loads. On opening day, Dr. Marden, the designer, "had some difficulty in operating his new crematory for burning the city's garbage." Three weeks later he still could not get his furnace to consume Milwaukee's wastes. The discouraged inventor, feeling the failure deeply, took his own life just days before Martin announced that, with the addition of another furnace, the "improved garbage crematory is a great success, every par-

[24] *Sentinel,* January 25, 1887. See also October 23, 1886. *Common Council Proceedings,* 1887, January 17, p. 366; March 21, p. 475; March 28, p. 492; April 13, p. 522; May 23, p. 69; June 6, pp. 93-94, 102; June 20, pp. 114, 116, 130; July 5, pp. 146-147; September 26, p. 311. See also the *Sentinel,* November 4, 1886, January 23, 25, 1887; June 8, 1887. See the company's incorporation papers in the Wisconsin State Historical Society archives, #M 530. It became the Forrestal Crematory & Garbage Company, reorganized in 1891 under the name Milwaukee Sanitary Company, and dissolved November 15, 1892, after it failed to get the city contract.

[25] *Sentinel,* May 20, 1887, p. 4; June 8, July 1, 1887, p. 3.

[26] *Sentinel,* June 11, 14, 1887.

[27] *Sentinel,* September 25, 1887, p. 4.

TABLE 4-1

Health and Garbage Expenditures, Milwaukee Health Department

	Health	Garbage	Total[a]	Garbage % of Total	Garbage Cost Index 1889=100	% Increase Of Garbage Cost Based On 1889=0
1869	2,507.75	—	2,507.75	—	—	—
1879	7,207.16	3,785.14	10,992.30	34.4	11.6	−88.4
1889 [Forrestal]	16,856.26	32,689.00	49,545.26	66	100	0
(1893) [Wis. Rendering Company]	65,610.20	74,666.59	140,276.79	53.2	228	+128
1899 [City]	34,085.07	61,883.21	95,968.28	64.5	189	+89
(1902) [Municipal Plant]	53,025.98	111,203.37	164,229.35	67.7	340	+240
1909	66,320.94	139,142.81	206,142.75	67.8	428	+328

[a] Excluding construction and sites.

ticle of matter . . . being entirely consumed without smell or smoke."[28]

If the furnaces seemed to be working well, the city collectors were not. Badgered by resident's complaints, the common council demanded that the health commissioner pay closer attention to collection. The members claimed that the garbage was left on the streets so long that it had "the audacity to attempt to remove itself, by crawling away, in the shape of active little worms."[29] Also, the crematory furnaces eventually showed themselves unable to dispose adequately of the daily garbage loads. Through bitter experience the health commissioner learned the headaches of collection and disposal and the frustrations of short-term private contracts. In 1889 the city, again contracting with the Forrestal company, tried to dispose of the city garbage by a new reduction method used in Buffalo, New York. Forrestal built a "Merz" rendering plant, which first placed the garbage in dryers and subjected it to high temperatures and then treated the residue with a chemical solution that extracted the oil. The Merz system produced grease and dry fertilizer that could be sold to make the plant economical.[30] The new method seemed more promising than cremation, and Health Commissioner Martin triumphantly travelled to the 1889 American Public Health Association

[28] The *Sentinel* followed the whole story of the search for Marden's body and the clues that he was still alive and had merely disappeared from home. See September 27, 29, October 5, 6. The newspaper finally announced on October 19, 1887: "At the time of his disappearance, Dr. Marden was interested with Wm. Forrestal in the city garbage crematory project, the success of which at that time was considered doubtful, although it has since proved to be successful. It was then thought that the apparent failure of the scheme had so preyed upon his mind that it brought on a fit of melancholia, from which he sought relief in death." See also *Sentinel*, November 2, 1887, p. 3.

[29] *Common Council Proceedings*, July 16, 1888, p. 168.

[30] For a description of the Merz process, see Edward Clark, "The Merz or Vienna System," *Public Health: Papers and Reports of the American Public Health Association* 17 (1891): 127-130. See also the *Sentinel*, June 24, 1889, p. 3 and July 13, 1889, p. 3.

meeting in Brooklyn, where garbage was a much-discussed subject, to proclaim that Milwaukee had, after fourteen years of trial and error, at last "happily answered" the garbage question.[31] But when a new health commissioner, Dr. U.O.B. Wingate, took office in 1890, he learned how premature his predecessor's optimism had been.

Three hundred "solid representative business men," residents in the vicinity of the Merz plant, greeted Dr. Wingate with a remonstrance against the "terrible odors" emanating from the furnaces, claiming that the stench "robbed us of our sleep and our meals."[32] As the West Side Anti-Stench Committee, they visited the institution, where many of them vomited from the vile odors, and threatened to take the city to court unless it abated the unhealthful nuisance.[33] Proponents of the motto "the stench must go" proved effective in commanding the attention of city authorities, and within weeks they convinced the crematory company to temporarily shut down.[34] The health commissioner once again faced the crisis of what to do with the more than forty tons of garbage produced in the city each day. He decided to consign it to the "bosom of Lake Michigan," admitting that modern technology had not yet solved the crucial health problem of how to dispose of large quantities of rapidly decaying organic refuse in American cities.[35] Loading the garbage onto lake boats offended persons who lived along the shore, who quickly let the health office know that they would not tolerate the nuisance any more than would the west-side residents. Echoing what must have been Dr. Wingate's sentiments, a city newspaper noted that "Like Banquo's ghost the garbage problem refuses to be downed."[36]

---

[31] Martin, pp. 63-64. [32] *Sentinel*, July 16, 1890, p. 1.

[33] *Sentinel*, July 30, 1890, p. 4; *Common Council Proceedings*, July 28, 1890, pp. 269-270.

[34] *Sentinel*, August 1, 1890, p. 1; August 3, 10, 1890; June 29, 1891, p. 4.

[35] *Sentinel*, August 6, 1890, p. 3; *Common Council Proceedings*, August 11, 1890, p. 289.

[36] *Sentinel*, August 18, 1890, p. 3.

In the fall of 1890, Dr. Wingate, hoping that the smells would be less offensive in the cooler weather, allowed the Merz plant to reopen.[37] The Anti-Stench Committee announced that it would begin court proceedings if Forrestal continued to run his garbage factory the following summer and accused the health commissioner of acting "as if he had a good block of stock in that most unsavory institution."[38] Through the winter of 1890-1891 the Merz plant rendered the city garbage while Wingate tried to figure out what to do next. Realizing that the location of the Forrestal plant in the middle of the city was "very unfortunate," he finally decided to cancel the Forrestal contract in anticipation of the west-side constituency's legal actions.[39] In February he presented the common council with the challenge of finding a better solution by June 1, when the factory would close.[40]

Also in February 1891, a group of businessmen, recognizing the financial potential of garbage, quietly organized the Wisconsin Rendering Company and laid plans to outstrip both Forrestal and the city in the race for lucrative solutions to the garbage dilemma.[41] In April this group of Milwaukee Democrats won the right from the state legislature to "purchase and use" land in Ozaukee county, north of Milwaukee, for garbage reduction. Because the city did not have the right of eminent domain outside its borders, it did not have the option that the legislature granted to

[37] *Sentinel*, September 13, 1890, p. 5.
[38] C. M. Cottrill quoted in the *Sentinel*, October 12, 1890, p. 3.
[39] MHD, *Annual Report*, 1891, pp. 28-29.
[40] *Common Council Proceedings*, February 24, 1891, p. 655. See also MHD, *Annual Report*, 1892, p. 68. Wingate also realized, and voiced for the first time, that a municipal plant would serve the city better than short-term contracts with private companies. He was ahead of the city in these thoughts. MHD, *Annual Report*, 1891, p. 29; *Journal*, July 13, 1891, p. 3. Meanwhile Forrestal was investigating alternative sites for his plant on the outskirts of the city. *Sentinel*, January 15, 1891, p. 3.
[41] Wisconsin Rendering Company incorporation papers, Wisconsin State Historical Society archives, #W 432, February 24, 1891. See also *Sentinel*, March 3, 1891, p. 3.

this new company. Nor did any other company enjoy the privilege of carrying Milwaukee garbage outside the city boundaries. Thus, before the city officials or the press understood the significance of its action, the Wisconsin Rendering Company had obtained exclusive rights to dispose of city garbage outside the city limits.[42]

Meanwhile the aldermen moved slowly in their search for garbage solutions. When the Forrestal plant closed on June 1, the city still had nothing to take its place. The council's inability to reach a decision apparently resulted from the competing influences among Democratic aldermen of the Forrestal Company and the newly formed Wisconsin Rendering Company. Accusations that certain aldermen were "pecuniarily interested" in one side or the other left the council at a stalemate and the garbage reeking in the yards and alleys of the city.[43] Residents complained about uncollected garbage; so Wingate decided to bury it in outlying farms. However, when the affected communities enjoined him from doing that, he resumed dumping it in Lake Michigan.[44] This led to complaints from residents and fishermen about the "chunks" of garbage in the lake— complaints that pushed Wingate to the breaking point. He threatened to leave the garbage "on the premises of citizens to rot," complaining, "I have done everything I possibly can to get this trouble settled."[45] He received no support from the common council or from city residents. To add to his troubles, Wingate got no clear message from his

[42] *Laws of Wisconsin*, 1891, p. 256. Passed April 17, 1891. See also *Journal*, April 20, 1891, p. 3 and *Sentinel*, April 25, 1891, p. 1. A discussion of the Wisconsin Rendering Company and the significance of eminent domain appears in Clay McShane, "Municipal Ownership and the Wisconsin Supreme Court 1870-1910," unpublished seminar paper, University of Wisconsin, 1970, pp. 13-14. Democratic alderman August Hanke and Democratic County Chairman H. J. Killilea exerted considerable influence over the state (Democratic) legislature to pass this enabling act. *Sentinel*, April 25, 1891, p. 1.

[43] See, for example, *Sentinel*, May 12, 13, 1891.

[44] *Journal*, June 19, July 8, 13, 31, 1891.

[45] Quoted in the *Sentinel*, July 31, 1891, p. 1.

fellow professionals about which method of garbage disposal was the most sanitary. Agreeing only on the health necessity of garbage collection and disposal, the American Public Health Association suggested that "each city should ... select [the method] found best for itself."[46] Looking at other cities, Wingate found that they, too, were experiencing garbage difficulties and were experimenting with various methods of disposal.

Milwaukee's crisis and Wingate's dilemma deepened in September 1891, when the council, amid bickering and accusations of "cliques" and payoffs, decided to accept none of the pending bids and to readvertise. Specifically, the Democrats in the council divided between Forrestal and Wisconsin Rendering Company, making a majority vote impossible.[47] The city had no choice but to continue dumping almost sixty tons of refuse into the lake each day. When Holton discovered—and vociferously called to everyone's attention—the garbage-polluted city water early in 1892, the aldermen, with a municipal election only a few months away, simply had to find a solution to their wretched sanitary problem.

Three possible solutions presented themselves in the early months of 1892. The city could issue municipal bonds and go into the garbage business itself; it could renew the contract with the old Forrestal company, which promised to upgrade its plant and even to move it from its central location; it could enter into a new contract with the Wisconsin Rendering Company, which could build its plant outside the city, where Milwaukeeans would not be bothered by bad fumes.

Dr. Wingate had earlier suggested that the city run its own garbage plant, but most aldermen and the public wor-

[46] "Report of the Committee on Disposal of Garbage and Refuse," *Public Health: Papers and Reports of the American Public Health Association* 29 (1903): 133. See similar statements in each *Public Health* from 1889, particularly the lengthy 1891 report in 17 (1891): 90-143.

[47] *Sentinel*, September 29, 1891. p. 1. See also editorial, September 30, 1891, p. 4.

ried about the "socialistic" implication of public owner-ship.[48] The easiest solution for the aldermen would have been to renew the Forrestal contract, since that company had bipartisan support in the common council. But aldermen seeking reelection worried considerably about the influence of the Anti-Stench Committee, which so adamantly refused to allow the old crematory to reopen.[49] The Wisconsin Rendering Company, with exclusive rights to dispose of city garbage outside the city limits, offered an advantage for aldermen seeking reelection: Milwaukee residents could not complain about odors arising from a plant fourteen miles from the city.[50] However, the Rendering Company bid considerably higher than its competitors, and the aldermen were reluctant to add to the city's fiscal problems in an election year.[51]

The process by which the council came to its decision about garbage disposal rested only partially on the merits of the various alternatives. Moving out of its deadlocked position in January 1892, the aldermen authorized an investigating committee to determine which method of disposal would be best for Milwaukee.[52] The committee toured American cities, viewing garbage works, and returned champions of the "improved" Merz system of disposal.[53]

[48] For discussion of the 1892 decision as an example of rejection of municipal ownership, see Bayard Still, *Milwaukee: The History of a City* (Madison: State Historical Society of Wisconsin, 1948), pp. 364-365; and David Thelen, *The New Citizenship: Origins of Progressivism in Wisconsin 1885-1900* (Columbia: University of Missouri Press, 1972), pp. 236-237.

[49] *Sentinel*, July 26, 1891, p. 3. The company offered to move their plant "wherever the council might direct," but their only options were within the city limits. *Sentinel*, July 28, 1891, p. 3.

[50] *Sentinel*, August 13, 1891, p. 3. As the *Sentinel*, reported, "it was worth, for the health of the city, almost any amount of money to have the garbage taken out of the city and county entirely."

[51] *Sentinel*, September 15, 1891, p. 4; September 16, 1891, p. 3. After submitting a bid to dispose of the garbage for $45,000 a year, the company revised it down to $35,000. However the Forrestal people submitted bids for $25,000. *Sentinel*, September 27, 1891, p. 7.

[52] *Sentinel*, January 23, 1892, p. 3; February 2, 3, 1892.

[53] *Sentinel*, February 18, 1892, p. 3; February 20, 1892, p. 1. The Merz

Initially its recommendation gave an advantage to the old Merz company, now headed by both Forrestal and H. A. Fleischmann of the Fleischmann Compressed Yeast Company. Just as the council was about to accept a bid from the Merz company, it received two new bids and stopped to consider them.[54] Fleischmann, worrying that he would lose his advantage, offered to take the entire council to St. Louis, where a Merz plant was successfully disposing of garbage.[55] The council, ignoring the fact that their committee had just toured the country and made its recommendations, decided to go. "The manner of the Common Council in handling the garbage question has been reckless from the start," commented one prominent official who did not travel to St. Louis, "but I must say that this last act is the most singular of all."[56] After accepting transportation, food, and drink from Fleischmann, the aldermen returned committed to Fleischmann's Merz process.[57] But they remained reluctant to allow the plant to operate within the city limits.

Although agreed on the value of the Merz process, council Democrats could not decide on the location of the plant, between the Forrestal advocates (inside the city) and the Rendering Company advocates (outside the city). The council was further stymied by some nonaligned Republicans and Democrats who remained open to other bids.[58] Caught before an election by citizens outraged at the state of affairs,

---

system had changed in these years, but it is impossible to determine exactly how. Advocates were so laudatory in their claims that specific variations are hard to find. The consensus of the Milwaukee investigators was that the Merz process was considerably better in 1892 than it had been in 1889.

[54] *Sentinel*, February 27, 1892, p. 2.

[55] *Daily News*, February 27, 1892, editorialized that any "free junket" appealed to the aldermen. See also *Sentinel*, February 28, 1892, p. 3.

[56] Quoted in the *Sentinel*, February 29, 1892, p. 3. The trip was the more incredible since St. Louis had adopted the Merz system only after visiting Forrestal's Milwaukee plant to see how it worked. See also *Sentinel*, March 5, 1892, p. 4.

[57] *Sentinel*, March 2, 1892, p. 3; *Journal*, March 2, 1892, p. 2.

[58] See, for example, the health committee hearings on the subject of

the aldermen finally allowed the private contractors to make their decision for them.

Behind closed doors and leaving no written record, the mayor and the health commissioner engineered a "reconciliation" between the Wisconsin Rendering Company and the Forrestal Merz people. As a result of the consolidation of the two companies, the city received a bid to erect a Merz plant on the Rendering Company's land in Ozaukee county.[59] Because the bid was higher than any previous bid from either company, rumor flew around the city that some aldermen had "profited from the transaction."[60] The health committee quickly recommended that the new bid be accepted, and the aldermen rallied around: ten Democrats who had previously voted against the Rendering Company supported the new combined company.[61] Only six aldermen voted against the new solution, described by the health commissioner as "one of the most important sanitary steps ever taken in our city."[62] Still hoping to reverse the decision by turning his opponents out of office in the election, Republican Alderman F. C. Lorenz accused the Democrats of underhanded tactics in obtaining the garbage contract. He exposed two Democratic "rings" whom the city would pay, he claimed, "an additional $45,000 a year" because of the consolidation.[63] Despite his charges, the "garbage cam-

garbage reported in the *Sentinel*, March 3, 1892, p. 3; *Daily News*, March 3, 1892, p. 1 and *Journal*, March 3, 1892, p. 3. The company representatives got into shouting matches during which Fleischmann called Gross a "patriarch from Jerusalem" and he returned by saying Fleischmann was "false as hell . . . you are full of yeast."

[59] *Sentinel*, March 4, 1892, p. 3; *Journal*, March 4, 1892, p. 4; and *Daily News*, March 4, 1892, p. 1. See also *Sentinel*, March 8, 1892, p. 1.

[60] *Daily News*, March 4, 1892, p. 1. See also *Daily News* editorial, March 5, 1892, p. 2.

[61] *Common Council Proceedings*, March 7, 1892. See the debate and the vote in the *Daily News*, *Journal*, and *Sentinel*, March 8, 1892. Immediately after the vote one of the losing contractors enjoined the city from executing the new contract, but the injunction did not stand. See the *Journal* analysis of the vote, March 9, 1892, p. 3.

[62] MHD, *Annual Report*, 1892, p. 42.

[63] See the correspondence between Alderman Lorenz and Health Com-

paign" of 1892 returned most of the incumbents to power and sustained the new contract. Of the six who voted against the contract, three were reelected (including Lorenz), two defeated, and one did not stand.[64] The voters felt obvious relief at having solved the garbage crisis and regarded accusations of personal gain as being of secondary concern.

Circumstantial evidence suggests some factors that might have led the Milwaukee Common Council to contract with the Merz Company to build a garbage disposal plant on Rendering Company land. Companies and people who had been adversaries united within days or even hours behind the merged company, suggesting that Lorenz was right about secret financial agreements. The 1892 contract cost the city more than the original contract with either Forrestal or the Rendering Company and thereby provided a financial cushion for such arrangements. It may be only a coincidence that simultaneously with these activities in Milwaukee the press revealed that the St. Paul, Minnesota, health commissioner was "financially interested" in Fleischmann's Garbage Collecting Company, which had a contract with that city.[65] Although there is no evidence that the Milwaukee health commissioner was financially implicated in the new Rendering Company, there is some indication that other officials were. A few stockholders of the Wisconsin Rendering Company can be identified. Charles Polacheck, Democratic assemblyman in the state legislature, was an active stockholder in the new company; he helped the company to obtain exclusive rights to dispose of the garbage outside the city limits.[66] Aldermen August Hanke (D), William F. Jordan (R), and A. J. Doelger, Jr. (R) probably had

---

missioner Wingate in the *Sentinel*, March 11, 1892, p. 5; March 18, p. 5; March 28, p. 5; March 31, p. 5. The quote is from the March 28 letter. See also the pre-election editorials in the *Sentinel*, April 2, 4, 1892, p. 4.

[64] *Journal*, April 6, 1892. I would like to thank William J. Orr, Jr., for his assistance in gathering local election results.

[65] *Sentinel*, April 29, 1892, p. 1.

[66] Polacheck was revealed to be a stockholder who was not above using bribery to convince others to his side in the second incident over the

a financial interest in the company; they were its most vocal advocates in the common council.[67] H. J. Killilea, the Democratic County Chairman, was the company's attorney; he guided the Rendering Company's rights of eminent domain through the state legislature. The Rendering Company members clearly used extralegal means to insure their political support. Their behavior in that direction became more explicit later in the company's history.

The Wisconsin Rendering Company began "relieving" the city of a "large amount of waste material" in August 1892.[68] Its price was high, but its service seemed worth it.[69] (See Table 4-1.) The company collected the garbage three

Wisconsin Rendering Company contract in 1897. See the discussion later in this chapter.

[67] These three aldermen were stockholders in the Rendering Company's antecedent, the Milwaukee Abbatoir Company, and remained loyal to, and in all probability financially connected with, Fred C. Gross, the Rendering Company's president. See the Milwaukee Abbatoir Company's incorporation papers, stockholders lists, and annual reports in the State Historical Society of Wisconsin archives, #M 2686. Some other nonpolitical figures can be identified as having an interest in both companies, including Fred C. Gross, Louis Boehme, and Joseph Schaaf.

The connection between the Milwaukee Abbatoir Company and the Wisconsin Rendering Company was suggested in the *Sentinel*, June 20, 1891, p. 4. We know, too, that the Milwaukee Common Council had previously signed contracts with companies in which aldermen had interests, as they did with the Milwaukee Coal Company, managed by Ald. Lorenz. *Sentinel*, July 12, 1891, p. 3. The *Sentinel* had accused contracting rings of operating to the detriment of the city on April 14, 1891, p. 3, listing William Forrestal as part of the "ring."

[68] The city considered fining the company when it was unable to open on its appointed day in June, but the new council decided not to strain relations so early in the five-year contract and extended the deadline until August 26, a date the company was able to meet. See the *Sentinel* passim during the summer, 1892, especially June 12, 26, 28, July 13, 14. A cholera scare raised anxiety levels and contributed to health department activity in the summer and fall of 1892. See the *Daily News*, August 25, 1892, p. 1 and the *Sentinel*, Augsut 31, September 2, 3, 9, 17, 19, 20, 29, and October 4, 1892.

[69] MHD, *Annual Report*, 1892, p. 35; *Sentinel*, October 5, 1892, p. 3. See also U.O.B. Wingate, "The Collection and Disposal of Animal and Vegetable Waste in Milwaukee," *Public Health: Papers and Reports of the American*

times a week, transported it to the docks, loaded it on barges, and disposed of it in the reduction plant fourteen miles north of the city. It used the improved Merz rendering process and sold the end product. If the plant smelled, Milwaukeeans were far enough away from it not to mind. Health Commissioner Wingate was delighted with the new sanitary service.

Two years later, when Republicans captured the city government and replaced Democratic Wingate with Republican Dr. Walter Kempster, the new health officer realized that garbage collectors were remiss in their duties. They neglected to collect some garbage and, more significant for the health officer, they did not destroy all the garbage they did collect, instead leaving much of it strewn about the beaches and floating in Lake Michigan. Kempster received 471 complaints against the company in the single month of June 1894. He complained to F. C. Gross, the president of the Rendering Company, about contract violations, but was unable to get any satisfaction before smallpox troubles erupted and diverted his attention. It was 1896 before Kempster could attend to the Rendering Company, and by then its negligence in collecting and disposing of the garbage had grown.[70]

In 1896 Commissioner Kempster, after initiating an investigation of conditions around the Merz plant, testified that "forty-eight tram car loads of garbage, dead animals and offal . . . [were] emptied into Lake Michigan at the foot of the [garbage] works . . . a large quantity of garbage, offal, etc., has been cast up upon the beach, and may there

---

Public Health Association 19 (1893): 49-51. See Table 4-1 for details on garbage and health costs in Milwaukee.

[70] To understand why Kempster could not attend to the garbage in 1894-1895, see the previous chapter or Judith W. Leavitt, "Politics and Public Health: The Smallpox Epidemic in Milwaukee 1894-1895," *Bulletin of the History of Medicine* 50 (1976): 553-568. See also *Sentinel,* July 11, 12, 1894, p. 3; *Evening Wisconsin,* August 7, 1894. Kempster's communication to the common council on Rendering Company inadequacy is quoted in MHD, *Annual Report,* 1897, p. 54.

be seen at this time."[71] City Engineer G. H. Benzenberg agreed with Kempster that the shoreline was "liberally supplied with garbage"; he noted that the plant capacity was sixty to seventy tons of garbage a day, whereas the city produced over eighty tons.[72] Isidor Ladoff, a health department chemist, testified that he had actually seen Wisconsin Rendering Company employees dumping several carloads of garbage into the water. Fishermen complained of the large quantities of garbage caught in their nets everyday. However, Alderman Doelger (R), chair of the council's health committee and foremost champion of the Rendering Company, visited the same site and later told newsmen: "We could not find as much as a wheelbarrow load of fresh or decayed garbage along the shore."[73] Rendering Company employees and officials and their aldermanic supporters insisted that they had dumped nothing in the lake except the small amounts that went astray during the unloading process at the water's edge.[74] The contradictory testimony led the council to suggest that the company may have been "derelict in the performance of its duties," but it did not recommend any punitive action. Instead, the aldermen held the health commissioner responsible for "not establishing such watch and vigilance as would have left no opportunity or temptation for the disposal of garbage in any [unsanitary] way."[75] The council's loyalty to the Rendering Company, among both Democrats and Republicans, was obvious, as was their continuing disdain for the health officer who had been so controversial during the smallpox epidemic.

[71] MHD, *Annual Report*, 1896, p. 58. See also *Common Council Proceedings*, August 17, 1896, pp. 424-426; *Sentinel*, August 18, 1896, p. 1.

[72] *Sentinel*, August 20, 1896, p. 3.

[73] Quoted in the *Sentinel*, August 21, 1896, p. 3. Other health committee aldermen agreed that Kempster's reports were "exaggerated."

[74] *Sentinel*, August 25, September 3, 1896.

[75] *Common Council Proceedings*, September 14, 1896, p. 480; *Sentinel*, September 13, 15, 1896; MHD, *Annual Report*, 1897, p. 61.

To prohibit the dumping of excess garbage in the lake, Kempster installed a lockbar on the company's tramway. He also increased health department monitoring of the company's collection.[76] But he directed most of his energy toward trying to get the council to favor a municipal plant over the private contractor, since he thought that the private system fostered methods that were "unbusinesslike, disgusting and unsanitary in every respect."[77] The open conflict between Kempster and the anti-Rendering Company advocates on the one hand and Doelger and the pro-Rendering Company advocates on the other continued for the next two years, during which garbage again emerged as a significant political issue.

The council, considering whether or not to renew the Rendering Company contract when it expired in 1897, divided between pro- and anti-Rendering Company factions and further split between those who advocated a municipal plant and those who opposed public ownership. Despite "strenuous efforts" of the Wisconsin Rendering Company, aldermen could not come to a decision.[78] They were further hampered by the national experts' inability to decide whether rendering or cremation was more sanitary. Rudolph Hering, who as head of the APHA committee on the disposal of garbage and refuse had thought about disposal more than any other individual, concluded: "No single system of disposal can be recommended as being the best under all conditions."[79] The council finally resorted

[76] *Common Council Proceedings*, September 28, 1896, p. 555.

[77] MHD, *Annual Report*, 1896, p. 60.

[78] *Common Council Proceedings*, November 23, 1896, pp. 640-642; MHD, *Annual Report*, 1897, pp. 68-70; *Sentinel*, November 24, December 5, 8, 15, 1896; *Sentinel*, April 13, 1897, p. 5. See also January 21, April 6, 29, 30 and May 1, 1897.

[79] "Report of the Committee on the Disposal of Garbage and Refuse," *Public Health: Papers and Reports* 22 (1897): 206-218. See also W. F. Morse, "The Collection and Disposal of the Refuse of Large Cities," *ibid.*, 20 (1895): 187-195; and the "Report of the Committee on the Disposal of Garbage and Refuse," *ibid.*, 20 (1895): 196-202.

to an old standby to determine which method Milwaukee should adopt: an investigating committee.[80] When the committee returned from its tour of American cities, it wrote a lengthy report comparing the various methods and recommended that Milwaukee change from the Merz to the Holthaus system, another rendering method that seemed less offensive.[81] The committee report was a blow to the Wisconsin Rendering Company, whose bid incorporated the Merz system, and it made that company redouble its efforts to draw members of the council away from the committee's recommendations. The Rendering Company had enough supporters in the council at least to delay adoption of measures contrary to its interests.[82] And indeed, in ensuing council votes, "the hand of the Wisconsin Rendering Company was visible" and "certain aldermen [were] determined to award the contract to the old company."[83]

Despite the Rendering Company's influence, in two crucial votes on July 6 and 7, 1897, the council administered what observers thought was "a final 'knockout' to the Wisconsin Rendering Company" by voting 31-11 and 34-8 in favor of locating a Holthaus disposal plant inside the city limits.[84] They then awarded the contract to the lowest bid-

[80] *Sentinel*, June 1, 3, 15, 16, 1897. City Engineer Benzenberg, Health Commissioner Kempster, and Aldermen Stevens, Renning, and Schranck toured the country. A rival committee set out on an independent tour. It was headed by Doelger and included Aldermen Ramsey, Thuering, and Rudolph and Dr. Nolte. *Sentinel*, June 10, 1897, p. 1.

[81] *Sentinel*, June 18, 1897; June 22, 23, 1897; *Common Council Proceedings*, June 21, 1897, pp. 185-192; MHD, *Annual Report*, 1897, pp. 102-121.

[82] During a council debate on whether the plant should be built inside or outside the city, "there was outspoken determination to save the Wisconsin Rendering Company" and 7 out of 13 aldermen on the joint committees voted for a site outside the city (i.e., for the Rendering Company). *Sentinel*, June 26, 1897, p. 2. Furthermore, delaying tactics in the council gave the company optimum time to influence the vote. *Sentinel*, June 29, 1897, p. 2.

[83] *Sentinel*, June 27, 1897, p. 5.

[84] *Sentinel*, July 7, 1897; *Common Council Proceedings*, July 6, 7, 1897, pp. 211-212, 255.

ders in those categories, Crilley and O'Donnell. The votes indicated that the Rendering Company's strength had dropped to six stalwarts: Doelger and five cohorts.[85]

But the Rendering Company did not give up. Although the *Sentinel* announced it had "exhausted every stratagem, defended every ditch, . . . and died in the trenches," the company and its allies remained active.[86] This became evident when Crilley and O'Donnell defaulted on their contract because they were unable to raise the necessary capital. Newspaper sources alleged that the Wisconsin Rendering Company persuaded capitalists to back out of their commitments to Crilley, offered Crilley $8,000 to abandon his share of the contract, and gave Emil Holthaus, the head of the rendering company, $3,100 to leave the city. A *Sentinel* reporter interviewed Holthaus at the railroad station, where the garbage magnate accused Attorney Killilea and President Gross of the Rendering Company of attempting to bribe him to kill the Crilley and O'Donnell contract.[87] The reach of the Rendering Company also became evident when the second bidders, Cooper & Burke, announced that they had a "business arrangement with the Wisconsin Rendering Company people for the lease of their property."[88]

In the face of the Rendering Company's "blatant act[s] of corporate arrogance" the common council took its first serious steps toward municipal ownership.[89] The *Sentinel*, wondering whether "the city garbage or the problem of

---

[85] *Sentinel*, July 8, 1897. The others were Berg, Eggert, Grootemaat, Niezorowski, and Ramsey.

[86] *Sentinel*, July 19, 1897, p. 5.

[87] Crilley's efforts to raise the money from people who went back on their original commitments are documented in the *Sentinel*, July 11, 13, 14, 15, 16, 17, 18, 21, 22, 23, 24, 25, 1897. The announcement of their inability to meet the contract specifications came July 28, 1897, p. 1, as did the bribery rumors. See also *Common Council Proceedings*, July 28, 1897, p. 291.

[88] *Sentinel*, July 31, 1897, p. 3.

[89] *Sentinel*, July 29, 1897, p. 1; *Common Council Proceedings*, July 30, 1897, pp. 297-300; August 2, 1897, p. 302. The phrase is Thelen's, p. 239.

how to dispose of it is the greater nuisance," supported an "experiment in municipal ownership" to get the city out of its recurring dilemma.[90] Realizing that the city was at the mercy of the Rendering Company, which, being the only group with the facilities to collect and dispose of city garbage, would benefit by default, Editor H. P. Myrick voiced "more than a suspicion that the emergency was made to order" and assured his readers that "it is not necessary to become a Socialist to enable a person to approve the suggestion of public ownership."[91]

The movement for municipal ownership received further encouragement when Charles Elkert, Republican alderman, announced that twice he had been "offered $300 to vote to give the garbage contract to the Wisconsin Rendering Company."[92] He formally charged Charles Polacheck (Democratic assemblyman and Rendering Company stockholder) in municipal court, which had Polacheck arrested and placed under bond.[93] Although the court eventually released Polacheck because of a state law granting immunity to members of the state legislature, Elkert's charges led to a Grand Jury investigation (which turned up no new evidence of corruption) and heightened the sensitivity of Milwaukeeans to the shortcomings of private contracting and to the strong-arm techniques of the Rendering Company. His accusations proved crucial in turning public opinion in favor of municipal ownership of a garbage plant.[94]

---

[90] *Sentinel*, July 31, 1897, p. 4.

[91] *Sentinel*, August 5, 1897, p. 4.

[92] *Sentinel*, August 10, 1897, p. 1.

[93] *Sentinel*, August 12, 1897, p. 1.

[94] Polacheck was arraigned in Police Court on August 12, 1897, charged with bribery, and scheduled for Municipal Court trial. *Sentinel*, August 13, 1897, p. 2. He was arraigned in Municipal Court on October 5 (*Sentinel*, October 6, 1897, p. 3), continued until the next session (*Sentinel*, November 12, 1897, p. 3), postponed (*Sentinel*, February 16, 17, 1898, p. 3), and finally dismissed on May 10, 1898 (*Sentinel*, May 10, 11, 1898, p. 5). The Grand Jury investigating aldermanic bribery met during October 1897, listened to testimony from city officials, contractors, and others, and con-

Still, the Rendering Company and its aldermanic supporters did not give up. Always six and sometimes as many as twelve council members continued to support the company's interests and stalled action on municipal bonds. Six aldermen even voted against the Grand Jury investigation of the bribery charges.[95]

As the time approached for the Rendering Company's contract to expire, Milwaukeeans, imagining garbage rotting on the hot streets, let their sentiments be known. The Municipal League and other citizen groups sponsored rallies during which people "declared strongly in favor of the municipal ownership." The "good government" citizens who attended these meetings denounced the "bold and high handed policy" of the Rendering Company advocates as "a deliberate insult to the patriotism and self-respect of every citizen of this great and prosperous city." Furthermore, they called the company's behavior "dastardly and corrupt . . . short-sighted and dishonest," and "a menace to the stability of government."[96] They sent up the cry for efficiency in city government.

The Rendering Company tried to disrupt these citizen gatherings, but succeeded only in further alienating Milwaukeeans.[97] After its contract expired on August 26, 1897,

---

cluded that it was "unable to find evidence of bribery." *Sentinel,* October 6, 7, 8, 9, 13, 14, 15, 1897. Quote is from October 29, 1897, p. 5. See also *Common Council Proceedings,* November 8, 1897, pp. 540-541.

[95] The "short list" of Rendering Company aldermen included Berg (Pop., 21), Doelger (R, 2), Eggert (R, 20), Niezorawski (D, 18), Ramsey (D, 3), and Grootemaat (R, 21). They were sometimes joined by Buestrin (R, 7), Dietrich (D, 19), Okershauser (R, 17), Weiher (D, 18), Hamizeski (D, 14), and Thiele (D, 9).

[96] Citizens of the twenty-first and thirteenth wards met on August 14, 1897. The Municipal League sponsored a mass rally on August 17, 1897. The quotes are from the resolutions adopted at the Municipal League meeting. *Sentinel,* August 18, 1897, p. 2. See also August 15, 1897, p. 1. For an examination of middle-class business reformers in this period, see Robert H. Wiebe, *The Search for Order, 1877-1920* (New York: Hill & Wang, 1967).

[97] *Sentinel,* August 18, 1897, pp. 1-2.

the company attempted to prevent the city trucks from collecting garbage.[98] Rendering Company officials also encouraged the outlying towns to enjoin the city from burying garbage on their lands.[99] The newspaper abounded with revelations about the company's continuing misbehavior. The council president denounced the "slimy, reptilious garbage corporation," and public sentiment seemed to turn wholly against the corporation.[100]

Nevertheless, the council, not yet ready for municipal ownership, voted in February 1898 to award the city contract to Cooper & Burke, a company that had already admitted its fiscal connection to the Rendering people.[101] The awarding of the contract was, in the *Sentinel's* words, "a complete victory for the Wisconsin Rendering Company."[102] In March 1898 the pro-municipal ownership mayor vetoed the contract, and the city again had no provision for garbage disposal.

Meanwhile Kempster devised a temporary system to dispose of the garbage by burying it within the city and in suburban towns that would allow it. Citizens did not lodge many complaints with the health department when it took over collection after the Rendering Company contract had expired, and Milwaukee saved approximately $2,000 a month over what it had paid the company.[103] The city's successful effort to cope, if only temporarily, with its wastes encour-

[98] *Sentinel*, August 27, 1897, p. 3.

[99] *Sentinel*, September 15, 18, 1897 and January 3, 1898, p. 3.

[100] Council President Baumgaertner was quoted in the *Sentinel*, February 1, 1898, p. 1.

[101] *Sentinel*, February 1, 1898, p. 1. See also February 24, 1898, p. 5.

[102] *Sentinel*, March 12, 1898, p. 1. *Common Council Proceedings*, March 11, 1898, pp. 778-780.

[103] MHD, *Annual Report*, 1897, pp. 152-153. There was a limitation on the city collection, however, which was necessary because the garbage was buried. No crockery, bottles, glass, tin cans, or like articles were collected, since those items could not be buried successfully without harming the soil. The Department of Public Works did offer to make such collections upon request. See the *Sentinel*, September 19, November 11, 1897.

aged many citizens to favor a permanent municipal solution to this most stubborn sanitary predicament.

The upcoming local election in 1898 finally brought the matter of a municipal garbage disposal plant to its conclusion. While newspapers and citizens groups charged the council members with "incompetency" and "stupidity" in their management of the garbage issue, all political parties came out in favor of a municipal plant.[104] Corporation activity had been so outrageous that even Alderman Doelger, seeking reelection, had to "bow to the inevitable and vote for the [bond] resolution."[105] But his action came too late, and on April 5 voters turned away the incumbent administration that had stood in the way of municipal ownership and replaced it with politicians who firmly supported it. Doelger and other Rendering Company faithfuls were defeated.[106]

Garbage—the subject of 35 discussions, 36 resolutions, 28 communications, and 21 committee reports in the common council between 1896 and 1898—helped to ensure a landslide Democratic victory and to bring Mayor David S. Rose to power. Ironically, this good-government Progressive issue brought into power an administration that ignored almost all of the good-government tenets. The garbage controversy and solution demonstrated that municipal

[104] *Sentinel*, February 16, 1898, p. 4. For other articles on the pre-election garbage controversy, see the *Sentinel*, February 2, 1898, p. 4; February 15, 1898, p. 1; February 25, 1898; March 2, 4, 12, 1898.

[105] The political campaign opened up further accusations against the Rendering Company when incumbents were denounced for their connections to the company. Andrezejewski (D, 14) was accused by his opponent as a "first-class boodler, [who] always has his hand behind his back and is receiving money from the large corporations in whose interests he votes." *Sentinel*, February 18, 1898. See also *Sentinel*, March 12, 18, 1898; Still, pp. 303-306. Of the six Rendering Company stalwarts, Berg, Ramsey, and Grootemaat did not stand for reelection; Doelger, Eggert, and Niezorawski were defeated.

[106] *Sentinel*, April 6, 1898, p. 1; Thelen, *The New Citizenship*, p. 238; Still, *Milwaukee*, p. 306.

ownership advocates and efficiency reformers both could benefit from working together, and garbage thus helped to prepare the way for the broad-based reform coalition that marked Milwaukee politics in the early twentieth century.[107]

After the 1898 election, the common council began the long task of working through legal blocks and municipal procedures to issue municipal bonds and to build the city garbage plant. Hoping one last time not to lose all its investment, the Rendering Company offered to lease its plant to the city, but the new council unanimously rejected its offer, deciding instead to locate the plant on Jones Island in the mouth of the Milwaukee River.[108] Freed at last from the corruptions and frustrations of the private contract system, the council decided to return to the cremation process, which seemed less offensive than reduction in a city location.[109] The city fought injunctions and debt ceilings

---

[107] For reaction to Rose's administration, see Still, pp. 307-315. The coalition between the Socialists and the middle-class business interests demonstrated here with regard to garbage also promoted general health reforms in Milwaukee.

[108] *Common Council Proceedings*, April 25, 1898, p. 16; May 9, 1898, p. 68; May 23, 1898, p. 120; June 6, 1898, p. 130; *Sentinel*, May 24, 1898, p. 3.

[109] Rudolph Hering, loathe to support any single method in his APHA reports, tended to favor cremation over reduction. Hering opposed reduction processes because frequently they were not economical, and because cities had to separate garbage from other refuse in order to use them. For more on this subject, see M. L. Davis, "The Disposal of Garbage," *JAMA* 31 (1898): 23-26; Rudolph Hering, "Report of the Committee on Disposal of Refuse Materials," *Public Health: Papers and Reports* 28 (1902): 21-28; "Report of the Committee on Disposal of Garbage and Refuse," *Public Health: Papers and Reports* 29 (1903): 129-133; C.-E.A. Winslow and P. Hansen, "Some Statistics of Garbage Disposal for the Larger American Cities in 1902," *Public Health: Papers and Reports* 29 (1903): 141-165. For opposing opinion see Charles V. Chapin, "The Collection and Disposal of Garbage in Providence, Rhode Island," *Public Health: Papers and Reports* 28 (1902): 46-50.

and finally began consuming its own garbage in March 1902.[110]

The new furnaces had the capacity to burn 120 tons of garbage daily. Collectors hauled the garbage across the harbor on scows and dumped it into a hoist that lifted the refuse into the drying pans at the top of the incinerator. Municipal ownership led not only to more efficient garbage collection and disposal by city employees, but also to increased responsibility and honesty on the part of the aldermen dealing with this problem. Garbage complaints decreased sharply during the first decade of the twentieth century, while accusations of corruption on garbage issues disappeared altogether. Municipal authorities, having put responsibility for garbage into the public sector, seemed determined to make it work.

Much more efficient than the Rendering Company's operations, the Jones Island plant nevertheless had its own limitations. During the harsh winter months, when the river was rough or clogged with ice, crossing the harbor was hazardous, and the health department was forced to resort to suburban burial. Furthermore, by 1906 Milwaukee produced more than 200 tons of garbage a day and severely taxed the plant's capacity. The new health commissioner in that year, Gerhard Bading, intent on efficiency in his

[110] For example, see the *Sentinel*, June 12, July 16, August 6, December 4, 1898; May 9, September 26, 1899; February 3, 1900. Still thought the plant opened in 1903, and Thelen and Goff also erroneously support that date. See Still, *Milwaukee*, p. 365; Thelen, *The New Citizenship*, p. 230; Charles Goff, "A History of Refuse and Garbage Disposal in Milwaukee County," chapter two of *Report to the Metropolitan Study Commission*, 1959. A typescript of this report is available at the City of Milwaukee Legislative Reference Bureau. Goff relied heavily on Rudolph Hering's 1907 report. None of these historians checked the health department sources, which clearly indicate that, in March 1902, 605 tons of Milwaukee garbage was cremated at the municipal plant. In April, it increased to 2,821 tons. MHD, *Annual Report*, 1902, p. 127.

Figure 17. Interior of the municipal incinerator plant,
c. 1909. Courtesy of Milwaukee County Historical
Society.

department, concluded that the city needed a new municipal incinerator in a more accessible location.[111]

Bading called in a national expert on garbage disposal,
Rudolph Hering, to help with the plans for the new plant.
Following his investigation, Hering reported to the Milwaukee Common Council in December 1907, applying his
vast technical knowledge to the local situation. He agreed
thoroughly with Bading that the shortcomings of the Jones
Island plant were capacity and location, and he disap-
proved of the department's practice of burying winter garbage in outlying lands.[112] He preferred cremation rather

[111] MHD, *Annual Report,* 1906, pp. 13-14.

[112] Hering had issued the yearly committee report in the APHA report
from 1889 to 1903. See also "Sewage and Solid Refuse Disposal," in
M. Ravenel, ed., *A Half Century of Public Health* (New York, 1921), pp.
181-196; *Collection and Disposal of Municipal Refuse* (1921) with Samuel A.
Greeley. Rudolph Hering was a civil engineer who had worked on surveys
for Prospect Park in Brooklyn and the extension of Fairmount Park in

than reduction for Milwaukee because the city had "barely enough grease in the garbage to make a reduction system profitable." Furthermore, Hering noted: "No complaints seem to have been made of odors emitted from the stack" of the incinerator. He recommended that Milwaukee build a new incinerator in a central location, and the city, to the delight of engineers around the country, agreed to follow Hering's suggestions.[113] *The Engineering News* voiced "pleasure" that, by consulting Hering, a major American city had realized that "garbage and refuse disposal [was] an engineering problem."[114]

The hiring of Hering was Milwaukee's first step in changing responsibility for garbage from the health department to the department of public works. The city began to consider sanitation a technical problem to be solved by technical experts, rather than a health problem to be solved by medical experts. The germ theory of disease supported this kind of thinking and provided the rationale for physicians

---

Philadelphia. He was city engineer in Philadelphia, 1876-1880, when he became interested in sewage problems in cities, a field in which little engineering had been done. He studied European sewage systems and wrote his *Report on European Sewerage Systems* (1881), which was the first comprehensive American study in the field. He supervised studies of water supply for Philadelphia and Chicago. In 1888 he moved to New York City, where he was consulting city engineer. It was estimated that he made sanitary reports for more than 250 American cities on water supply, sewage disposal, and refuse disposal. He was president of the American Public Health Association in 1912-1913. See the *Dictionary of American Biography*, pp. 576-577. His report to the Milwaukee Common Council, "Report on Garbage Disposal," is reprinted in MHD, *Annual Report*, 1907, pp. 140-188.

[113] Hering, "Report to the Common Council," pp. 145, 159, 160-161, 163, 188. Hering's description of Milwaukee is amazingly astute for the length of time he spent in the city. He described the population ward by ward and analyzed the garbage by characteristic population of various wards. He found that the quantity of garbage, rubbish, and waste coal was usually much greater in the well-to-do districts than in those occupied chiefly by a laboring population (p. 156).

[114] Quoted from *Engineering News* 59 (1908) in the MHD, *Annual Report*, 1907, p. 13.

to want to switch the health department emphasis from cleaning the environment to pursuing specific diseases. Of course, as Bading realized, Milwaukee had already virtually "solved the waste disposal problem."[115]

The new municipal incinerator, which efficiently burned all garbage, ashes, rubbish, and street sweepings produced in the city, opened in April 1910. This successful plant continued to serve Milwaukee until 1955.[116] In 1911 the city consolidated the tasks of collecting and disposing of all municipal wastes under the single authority of the department of public works.[117] Municipal ownership, the return to cremation, the city location, and increasing reliance on experts had all combined to make garbage an engineering rather than a health problem in Milwaukee.

Despite nineteenth-century certainty about the connection between dirt and disease, the extent to which garbage posed a direct threat to the health of Milwaukeeans remains unclear. With the exception of the 1892 crisis over garbage-contaminated water, it is not possible to blame garbage for creating any specific health problems. Simple correlations cannot be made between sanitary garbage disposal and improving health statistics. During the very years when Milwaukee grappled with sanitation crises and solved them, city mortality rates indeed decreased. But since those very years—from the 1890s to the 1910s—were also years of

---

[115] MHD, *Annual Report*, 1909, p. 13.

[116] MHD, *Annual Report*, 1910, p. 121. It was called Erie Street No. 1. The Erie Street No. 2 was built in 1930 to help with additional loads. The Erie Street No. 1 began its fires on April 14, 1910; the Jones Island plant closed on May 11, 1910. All city garbage was disposed of at the new Erie Street plant beginning May 12, 1910. See the MHD, *Annual Report*, 1910, p. 116; Goff, "History of Refuse."

[117] Such efficiency studies and recommendations were made for many of Milwaukee's municipal agencies under the Socialist regime of 1910-1912. The Bureau of Economy and Efficiency made the studies under the direction of John R. Commons. See the Bureau of Economy and Efficiency, Bulletin No. 5, "The Refuse Incinerator" (1911) and Bulletin No. 10, "Garbage Collection" (1912).

great change in so many other aspects of city life, it is impossible to prove a direct connection with any single environmental concern. Vast quantities of rotting organic wastes, given the proper temperature and aided by insect vectors, could have produced a serious menace to the water supply and to the city's health. Diarrheal diseases, including typhoid fever, might have spread through garbage-contaminated water supplies, although raw sewage, which contained human excreta, would have been the most likely perpetrator of any major health disasters. Garbage exacerbated health-related problems, but probably did not itself cause the major hazards. The quest for sanitary garbage disposal thus tells us more about the process of health reform than about the causes of improved health statistics.[118]

[118] For discussion of the role of hygienic reforms in improving mortality, see McKeown, *The Role of Medicine*, pp. 121-124. A study of sewage or water instead of garbage would have provided a more dramatic illustration of the connections between sanitary reform and improving mortality. See Edward Meeker, "The Improving Health of the United States," who connects declining typhoid fever death rates with improving urban water supplies.

CHAPTER FIVE

The Politics of Health Reform:
Milk

"*Milk, as secreted* by the healthy cow, is the purest food
we know," said Health Commissioner F. A. Kraft in 1911.
But, Kraft continued in his remarks during a health department campaign to improve the city's milk supply, "It
is man who soils and befouls and contaminates."[1] The health
official well knew that much of the milk that Milwaukeeans
drank harbored hidden dangers. Often the innocent-look-
ing "lacteal fluid" contained disease-causing bacteria from
a sick cow or from contamination along the route from the
cow to the breakfast table.

Nineteenth-century Milwaukeeans received their milk
from both urban and rural cows. As the city increased in
size, milk from both sources became hazardous to the health
of milk drinkers. When urban land grew congested, dairy
owners had no place to pasture their herds; thus they adopted
the practice of keeping them penned inside barns for most
of the year. Furthermore, in a city famous for its breweries,
distillery slops—alcoholic waste products—provided an
economical staple of the urban cow diet.[2] Milk produced

---

[1] *The Healthologist*, Milwaukee Health Department Bulletin, October,
1911, p. 1.

[2] A similar situation existed in New York City. For efforts of that city
to deal with the problem, see Norman Shaftel, "A History of the Purification of Milk in New York or How Now Brown Cow," *New York State*

156

by the slop-fed animals, who stood, slept, and gave milk in crowded stables amid their own excrement, became the late-nineteenth-century focus of the health officers' attacks on the quality of milk in the city.

Rural milk was probably not too much safer than city milk. As the city expanded geographically, dairies moved farther and farther from the people whom they supplied. Although farmers did not feed their cows distillery swill exclusively and gave them the freedom of the pasture, country milk also harbored disease. Many rural cows carried and transmitted tuberculosis and other diseases to the unsuspecting drinker. Cows from the pasture returned to the barn for milking with udders and flanks soiled from the earth and their own excrement; this filth often fell into the open milk pails. Dairy employees took few precautions with their hands and clothes. Milk cans, frequently unwashed, stood unattended in the barn during milking, and then waited by the roadside for transport to the city. Wholesalers hauled the milk over dusty roads or loaded it onto unrefrigerated trains, and transporters often relieved their thirst by dipping into the milk cans.[3] Both urban and rural milk thus had many opportunities to become contaminated

---

Journal of Medicine 58 (1958): 911-928. See also Robert Hartley, *An Historical Scientific and Practical Essay on Milk as an Article of Human Sustenance; with the Consideration of the Effects Consequent upon the Present Unnatural Methods of Producing it for the Supply of Large Cities* (New York, 1842); James Flexner, "The Battle for Pure Milk in New York City" in New York Milk Commission, *Is Loose Milk a Health Hazard?* (New York: Health Department, 1931); and John Duffy, *Public Health in New York City*, I, pp. 420-439.

[3] For a detailed description of the process of getting country milk to market in Milwaukee, see Peter J. Weber, "The Municipal Milk Supply," *The Healthologist*, October 1911, pp. 4-13. See also Charles Harrington, "The Practical Side of the Question of Milk Supply," *Pediatrics* 16 (1904): 209; George M. Kober, "Milk Sediments or Dirty Milk in Relation to Disease," *JAMA* 49 (1907): 1091; and J. Cheston Morris, "The Milk Supply of Our Large Cities: The Extent of Adulteration and its Consequences: Methods of Prevention," *Public Health: Papers and Reports* 10 (1884): 246-252.

Figure 18. Driver waits while customer pours milk in his own container, Milwaukee, 1900. Courtesy of Milwaukee Public Library.

from warm temperatures, unsanitary handling, and human contact.

Marketing procedures added to the dangers of drinking milk. Until well into the twentieth century most urbanites obtained their milk by "open dipping" from street vendors or local grocers. The loose handling of the milk provided numerous occasions for contamination to enter the milk supply. Because most consumers had no way of chilling the milk that they bought, bacteria had a further chance to flourish. Milwaukee's milk, instead of being the "purest food," undoubtedly caused many cases of tuberculosis, diphtheria, and childhood diarrheal diseases, increasing the number of deaths that plagued urbanites in the late nineteenth and early twentieth centuries.[4]

[4] For some contemporary expositions of the problem of disease in milk, see, M. J. Rosenau, ed., *Milk and its Relation to Public Health* (Washington: U.S. Treasury Department Public Health Service, Hygienic Laboratory

Milwaukee's first health officer, James Johnson, thought there was "little cause for complaint" about the city's milk supply.[5] Johnson did not devote much time or effort to food control because he felt it was secondary to the larger concerns of infectious diseases and sanitation. However, his successor, Orlando Wight, who linked "poor and unhealthy milk" to "sickness and . . . death among the children of the city," made milk one of his major crusades.[6] Wight vowed increasing and continual vigilance, especially after realizing from the statistics collected in his office that over sixty percent of Milwaukee's mortality struck children under the age of five. But when he proposed to license milk dealers and make them accountable for dispensing impure or unhealthy milk, Wight met immediate opposition. The city attorney ruled that the health commissioner had no power to regulate the sale of milk because milk dealers were not hawkers or peddlers.[7] Wight did manage to ease through an ordinance prohibiting the sale of impure milk in the city, but because it included no enforcement provisions, most dairies ignored it.[8]

Wight refused to be daunted by legalisms. He compiled a list of milk producers in the city, noting the condition of their barns and animal feed, and then posted it, so con-

---

Bulletin No. 56, 1909); John Spargo, *The Common Sense of the Milk Question* (New York: MacMillan Company, 1910); Horatio Newton Parker, *City Milk Supply* (New York: McGraw Hill, 1917); William G. Savage, *Milk and the Public Health* (London: Macmillan and Company, 1912); M. J. Rosenau, *The Milk Question* (Boston: Houghton Mifflin Co., 1912), pp. 92-112. See also numerous articles appearing in medical journals, i.e.: Ernest Wende, "City Milk Routes and their Relation to Infectious Diseases," *JAMA* 34 (1900): 150-153; John Mohler, "Various Diseases and Conditions of Cattle That May Affect the Quality of the Milk Supply," *JAMA* 49 (1907): 1089-1091.

[5] MHD, *Annual Report*, 1875, pp. 14-16; MHD, *Annual Report*, 1877, p. 46.

[6] MHD, *Annual Report*, 1879, p. 11.

[7] Milwaukee *Sentinel*, July 9, 1878.

[8] *Ordinances of the City of Milwaukee 1878*. The health ordinances were reproduced in full in the MHD, *Annual Report*, 1878, p. 57.

sumers could learn the conditions under which their milk was produced. He hoped in his way that people would stop buying milk from unsanitary dairies and that the milk trade could regulate itself. "If anybody's 'pure' milk business is thereby injured," Wight concluded, "it will not be the fault of this office."[9]

Wight's investigations carried him into 227 milk stables in 1878 and 1879 and further substantiated in his mind the need for city control over this important food. Armed with his data, he made a second appeal to the common council to allow regulations of the milk sold in the city and thus to relieve the burden on the "little ones." Wight emphasized the economic consequences of unsanitary milk, thinking the council members would thus be inclined to listen. Milwaukeeans purchased approximately 17,000 quarts of milk each day, he calculated. Each quart sold for five cents, which meant that the city expended $850 a day on milk, or $310,505.50 each year. Wight concluded that the city suffered an economic loss whenever milk was adulterated by skimming or watering, made unwholesome by filthy stables, or taken from diseased cows. Not only was tainted milk economically unsound, but the bad business habits often led to tragedy: "[N]one but Heaven could measure the sickness and suffering, or record the deaths thereby caused among children in the various households of the city. Against such considerations no fictitious 'personal liberty' to be nasty, no spurious 'freedom of trade' to supply the people with adulterated and polluted food, can stand a moment."[10]

Wight uncovered conditions in the cow stables of Milwaukee similar to what New York City's health reformers had found earlier. Of the approximately 3,000 cows that he inspected, 107 never left their stables, 2,142 remained confined nine months of the year, and one-quarter ate

[9] Dr. Wight was quoted in the Milwaukee *Sentinel*, July 20, 1878.
[10] MHD, *Annual Report*, 1879, p. 79.

160

nothing but distillery slops. Many of Milwaukee's cows lived in stables loaded with "putrescent vapors." According to the health commissioner: "The liquid manure has run through the floors, saturated the earth beneath, undergoing putrefactive fermentation and generating poisonous organic vapors." Only 52 stables inhabited by 713 cows produced milk that merited approval for human consumption. "It is obvious," Wight concluded, "that the business of supplying milk to the families of this city ought to be put under some restraint."[11]

Ideally Wight wanted to control the entire process of milk production, from the conditions of the cow barn to the handling and transportation of the milk, both inside and outside the city. He sought powers to require pasturage for all cows, well-drained and watertight stables, daily removal of dung, ventilation of cow stalls, sanitary handling of milking utensils, and regulation of the cow's diet. His comprehensive view of how the city should attack its milk problem provoked a strong reaction against increasing health department powers. The *Daily News* led a successful campaign against enlarging city milk controls. Maintaining that the proposed ordinance would "utterly ruin" business, this working-class newspaper ridiculed the provision of required pasturage, pointing out that urban milk producers could "find no pasture in the city, while their cows would be of no use to them in the country, there having been as yet invented no device by which milking can be done by telephone."[12]

The dairy workers' protest went beyond making fun of the "wind-bag" health commissioner. Approximately two hundred milkmen joined together to "resist this attack on

[11] *Ibid.*, pp. 80, 82. See accounts of Wight's appearance before the council as well as a copy of his communication to that body in *Evening Wisconsin*, June 7, 1879; *Milwaukee Sunday Telegraph*, June 8, 15, 1879; *Daily News*, June 10, 1879, and MHD, *Annual Report*, 1879, pp. 78-104.

[12] *Daily News*, June 13, 1879.

their liberties."[13] Their actions revealed the economic divisions that characterized milk politics in Milwaukee. An estimated four hundred dairies inside and outside the city supplied Milwaukee with milk. Most of these enterprises were one-person operations consisting of only a few cows and limited physical facilities. The small producers, largely German and Polish, opposed any health reforms that might destroy their marginal profits; the larger producers, perhaps realizing that much of their competition could be eliminated through judicious support of sanitary improvements, increasingly supported health department efforts to control milk production. In the eyes of the newspapers, the "best dealers," who "assent to the proposed restrictions," thus countered the "blatherskite milkmen," who "raise a row against the city regulations."[14]

At the common council hearings on Wight's proposed ordinance, the division between the milk producers became increasingly visible. The German and Polish milk dealers denounced the proposed regulations as a "fraud . . . in the interest of the larger milk dealers who wished to drive their weaker brethren out of the business." These dealers, who won the support of many aldermen, petitioned the council to defeat Wight's ordinance because it "would entail great hardship and suffering" on classes "entirely dependent on their business for a livelihood."[15]

Wight answered his critics in a patriotic July 4 letter to Milwaukee citizens, defending the cause of 2,500 babies against the "militant milkmen" and their aldermanic supporters. He particularly criticized the "ward politicians,"

[13] *Daily News*, June 15, 1879. Newspapers and contemporary records frequently referred to the milk dealers as "milkmen" or "cowmen," although many women pursued this trade in Milwaukee.

[14] *Milwaukee Sunday Telegraph*, June 22, 1879. See also *Evening Wisconsin* June 21, 1879; *Sentinel*, July 9, 1879; and MHD, *Annual Report*, 1879, p. 90.

[15] *Daily News*, June 22, 1879; Milwaukee *Sentinel*, July 4, 1879. See also *Daily News*, July 4, 1879.

who, he implied, had aroused the milk dealers to action. He denied any favoritism toward the large milk producers and testified to the cleanliness of many small dairies. "I have no desire to hurt anyone," he protested. "With ignorance and prejudice I shall be patient, but I shall not be very tolerant with the 'cussedness' that regards the loss of one hundred votes as a much greater calamity than one hundred dead babies."[16] The health commissioner's language and message did not endear him to the municipal legislators, who rapidly put an end to the controversy by indefinitely tabling the ordinance.[17]

Wight's clash with the aldermen in his 1879 attempt to obtain effective milk controls raised issues that plagued future debates over the regulation of food. Did the city have a right to regulate business? What authority did the city have over enterprises located outside the city limits? To what extent were activities of the health department harmful to the working people and helpful to the moneyed interests? Whose interests did the health commissioner represent?

Important in explaining Wight's failure to achieve his goals in 1879 was the state of medical knowledge about the connections between milk and disease. Although some physicians, like Wight, posited a causal relationship between tainted milk and childhood morbidity and mortality, the precise nature of what rendered the milk hazardous was not yet understood. Many physicians admitted the difficulty of defending specific reforms before they had demonstrated specific causal connections. In the face of this medical indecision, the political and economic divisions became particularly significant. Fear of lost votes, defecting constituents, and adverse economic ramifications moved the

[16] Quoted in the *Sentinel*, July 5, 1879.

[17] *Sentinel*, July 8, 11, 22, 1879; *Evening Wisconsin*, July 9, 10, 29 and August 4, 1879; and *Sunday Telegraph*, July 13, 1879; MHD, *Annual Report*, 1879, pp. 93-94. See also *Daily News*, August 5, 1879; *Sentinel*, August 9, 1879.

aldermen to oppose an increase in health department activity. The public supported their opposition because, not being able to distinguish between pure and tainted milk, they were eager to keep the price of milk down. Although Wight won some support from the more affluent dairy producers and the sympathy of many fellow physicians, he failed to gather a coalition with enough political influence to compete with those who opposed him.

Wight's successor, Robert Martin, continued to pursue milk regulation. In December 1882 he invited a *Sentinel* reporter to join him in visiting some urban cow stables. They found—and the *Sentinel* subsequently publicized— foul conditions in the dairies that produced Milwaukee's milk. In one barn: "Fourteen cows huddled together in a miserable hovel, embedded in reeking filth. . . . The poor creatures are tied up in the fall, and not let out for exercise or a breath of fresh air until the following spring."[18] Vile odors assailed their nostrils as the physician and the reporter made their way from stable to stable. Citizens read about the "Cows by the score, with tails rotting off by the disgusting nastiness in which they drag on their lives." The health commissioner revealed conditions of the cows' lives in graphic detail: "Pools of noxious water and refuse covered the floor, which was soft and soaked with foul liquid. The eight cows stood closely tied, their sides touching. None were lying down, and it did not look as though they could . . . although their quarters and flanks, that were thickly coated with manure, showed that they did change their positions. Their hides were damp with sweat, and several were panting laboriously."[19] The *Sentinel* reporter uncovered similar conditions in barns all around the city, and his exposé made lively newspaper copy during the Christmas season.

[18] *Sentinel*, December 28, 1882, p. 2.
[19] *Sentinel*, December 30, 1882.

To strengthen the impact of his revelation, the *Sentinel* reporter interviewed physicians about the health implications of the foul conditions. Most agreed readily that milk produced under such conditions was a significant cause of disease and death among the city's children. Dr. Solon Marks, head of the State Board of Health, called the use of swill milk "injurious . . . producing [in children] irritation of the stomach, diarrhoea, and inducing nervous troubles." Dr. Ernst Kramer, a former critic of the health department, applauded Martin's efforts to regulate milk production. He admitted that he could not directly trace disease in children to the milk supply, but he supported the potential value of Martin's proposed regulation. Other doctors agreed that, although it was logical to assume a connection between bad milk and high infant mortality, the two could not be linked for certain. At least one physician interviewed believed that swill milk was safe. "I have eaten the beef from distillery

Figure 19. Inside a Milwaukee milk barn, showing skeleton remains, 1891. Courtesy of City of Milwaukee Health Department.

165

swill fed cows and have drank their milk," said Dr. James
M. Allen, "and it never did me any harm."[20]

In order to rid the city of swill milk, produced by cows
fed on brewery slops, and to force clean physical conditions
in the cow barns, Martin submitted an ordinance to outlaw
the sale of impure milk and regulate dealers through licensing.[21] He received some public support for his proposal
because of the publicity about foul cowbarns, but even the
Sentinel refused to give Martin all the help he needed. "There
is no need for a general scare on the subject of milk,"
concluded the *Sentinel* editor. "The present need is for persons who have invalids and infants in charge to visit and
personally inspect the dairies." The *Sentinel* believed that,
despite the wretched conditions, most Milwaukee milk came
from clean dairies and well-fed cows.[22]

The dairy producers organized for battle against Martin's proposed ordinance and threatened to raise milk prices
if forced to modify their production methods. In the process they admitted that they could not produce unwatered,
non-swill milk for the current market price of five cents a
quart. They claimed "the personal right to feed their cows
whatever swills they may choose and to keep their stables
as foul as their tastes and interests demand."[23] The dairy
interests again convinced the council members that the public
body had no right to infringe on business practice or to
regulate a trade practiced outside the city: the result was
another stalemate and no milk ordinance.

Thwarted by the common council, Health Commissioner

[20] *Sentinel*, January 2, 1883. The physicians interviewed were for the
most part among the well-known names in Milwaukee. They included
Drs. Nicolas Senn, Solon Marks, Ernst Kramer, Herman Nauman, A. J.
Hare, Lewis Sherman, William Fox, James M. Allen, and William Thorndike. See also January 4, 1883.

[21] *Common Council Proceedings*, January 8, 1883, p. 293.

[22] *Sentinel*, January 5, 1883, p. 4.

[23] *Sentinel*, January 5, 1883, p. 4; *Sentinel*, January 23, 1883. See also
January 25, February 3, 1883.

Martin changed his tactics. Instead of trying to affect public opinion in the newspapers or trying to convince aldermen to support a milk ordinance, he began using his nuisance powers slowly to clean up the physical conditions of the urban cowbarns, at the same time trying to convince individual dealers to alter their feeding patterns. Martin summoned twenty-five dealers who maintained filthy premises to appear at the health office to explain why they did not end these conditions. He accused the dealers of keeping hogs on their property and locking their cows in unventilated, filthy barns. The results of the conferences with the dealers pleased Martin: "All . . . promised to clean up their premises at once, and some even went so far as to say they would feed no more swill to their cattle."[24] Martin knew that success achieved under these nuisance auspices would be only temporary and that most of Milwaukee's milk, produced outside the city limits, could not be controlled in this way. Thus he continued to seek legal powers to regulate the city milk supply, while milk dealers continued to assert their personal rights to decide the conditions under which they would produce milk.

In 1887 Martin introduced another milk ordinance for council consideration. This one called for an office of milk inspector within the health department and for licensure of all milk vendors. Prerequisites for licensing would include no slop feeding and clean cowsheds. Again Martin found the traditional supporters of the health department—the *Sentinel*, physicians, and some municipal organizations—less helpful than he wished. The *Sentinel* worried that sanitarians overemphasized the harmful qualities of swill milk. "[T]here has been no sufficient collection of evidence to place the noxious character of this food beyond all question," the newspaper asserted.[25] Until it saw such

24 *Sentinel*, January 14, 1885, p. 3.
25 *Sentinel*, June 14, 1887.

conclusive evidence, the *Sentinel* remained unenthusiastic and only moderately supportive.

To strengthen his crusade, Martin collected and analyzed samples of milk sold in the city, demonstrating that eight out of ten vials of the "lacteal fluid" had been watered and at least four had come from swill-fed cows. Although he thought this evidence proved the need for public regulation, the *Sentinel* opined that "on the whole, the showing . . . is better than most persons would have expected."[26] The editors, however, did concede that the milk supply could be improved, and they halfheartedly supported Martin's efforts at regulation.

Physicians generally supported Martin's contention that swill or adulterated milk could endanger children's health. In addition, the assistant director of the Wisconsin Agricultural Experimental Station in Madison aided Martin by testifying that brewers' wastes should not be the sole diet of cows. The Humane Society argued that children and cows needed public protection against milk adulterators.[27] As experts increasingly agreed upon the need for some regulation, the alliance between the larger dealers and the health department strengthened. The major producers, who had some extra means, worked to modify the harshest parts of the milk ordinance and then supported it against the protests of their weaker and poorer colleagues.[28] In a compromise move Martin agreed to remove his total exclusion of slop-feeding and to substitute a provision requiring that cows not be "fed on distillery slop or swill or garbage in such quantities, unmixed with other food, as to impair the quality of milk or make the same unhealthful."[29] The revised ordinance did not specify the allowable proportions.

The smaller dealers vowed to resist any regulatory or-

[26] *Sentinel*, July 27, 28, 1887.
[27] *Sentinel*, August 7, 1887; February 24, 1888.
[28] *Sentinel*, August 6, September 9, 1887.
[29] *Sentinel*, August 17, 1887. See also *Common Council Proceedings*, August 15, 1887, p. 221.

dinance and refused to register for milk licenses.[30] But their political influence proved as limited as their economic resources, and in December 1887 the council passed a milk ordinance, Milwaukee's first comprehensive regulatory legislation governing food.[31]

Health Commissioner Martin, delighted with his successful ordinance even though its compromise version did not give him all the powers that he felt he needed, soon realized that passage of a law marked only the beginning of a difficult enforcement battle. He knew that many of the small city dairies, maintaining three to five cows, could not survive the new regulations. "Many of our vendors will be driven out of business," Martin admitted, "and I don't much care if they are. Cows properly belong in the country."[32] Martin's intolerant attitude toward the marginal dealers angered the milk producers and furthered their determination to resist his encroachment. If he had been politically prudent, he would have confided his willingness to eliminate some dealers only to friends; instead he publicly advocated monopolies to obtain more uniform and regulated milk production, a policy that frightened even some of the larger dealers.[33]

In January 1888, when the date for applying for milk licenses arrived, most milkmen stayed away from the health office. Even the conciliatory association of the larger dealers resisted the measures when its members realized that the ordinance came down more strongly on urban cowmen than on suburban dealers. The urban vendors, large and small, presented a united front against the "barn clause," which stipulated that city barns had to be whitewashed an-

[30] *Sentinel*, August 22, 1887.
[31] *Sentinel*, October 7, 1887, p. 7; *Common Council Proceedings*, December 19, 1887, p. 417. The vote was 29-1, Alderman Dunck voting no. See also *Sentinel*, December 20, 24, 1887.
[32] *Sentinel*, December 6, 1887, p. 3. See also *Sentinel*, December 24, 1887; February 3, 1888.
[33] *Sentinel*, January 5, 1888, p. 3.

nually and that cows had to be turned out for exercise on fair days, provisions they saw as discriminatory. They also objected to the requirement that all milk sold in the city contain twelve percent solids.[34] In the end they managed to introduce a substitute ordinance.[35]

For months the milk debate raged in Milwaukee, sorely testing the health commissioner's authority. Some milkmen supported Martin, whereas most defied him. Martin nevertheless continued to expose the unsanitary conditions in city dairies. He uncovered one instance of adulteration with annetto dye, but when he brought the offender to court, the judge ruled that Martin had to prove the unhealthfulness of the coloring substance in order to prosecute.[36] Meanwhile the dairymen successfully resisted the health department by refusing to register for licenses. By February 27 only an estimated one-seventh of the total number of dealers had applied for licenses.[37] In an effort to save face, Commissioner Martin extended the deadline for applications, but he was "doomed to disappointment."[38] By April 17 Martin estimated that not even one-fourth of the cow owners were in compliance.[39] The rest were working busily behind the scenes for an amended ordinance. After their first efforts failed, the milk dealers succeeded in August 1888 in getting the council to pass a revised ordinance that deleted most of Wight's original requirements. The milk dealers seemed prepared to comply with the new bill's required registration, the substitute for licensing, although registration would not provide Martin with regulatory

[34] *Sentinel*, January 9, 1888, p. 3; January 19, 1888, p. 3; January 20, 1888, p. 3.

[35] *Common Council Proceedings*, January 16, 1888, p. 473. The substitute ordinance was introduced by Alderman Dunck, whose vote had been the only dissent heard when the original measure passed in 1887. He was from the ninth ward.

[36] *Sentinel*, September 12, 1888, p. 3.

[37] *Sentinel*, February 27, 1888.

[38] *Sentinel*, March 2, 1888.

[39] *Sentinel*, April 17, 1888, p. 3.

power.[40] The health department still could control conditions in cow stables only insofar as they were public nuisances.

Although registering did not require passing health department inspection, milk dealers remained reluctant to sign their names in the department registry. Out of an estimated 400 dealers, 230 of whom were urban residents, "only a few" had put their names on the books by September 19, 1888, one day before the deadline.[41] This time Martin brooked no disobedience and asserted his power over the recalcitrant milkmen by arresting one of them, a small dealer named John Weimer, for selling milk in the city without having registered to do so. Martin thus called on the courts to settle the issue of power and rights.

Some of the milk producers who had not signed the registry hastened to Weimer's aid, pledging financial and moral support.[42] Others refused to undermine an ordinance that many of them had devised. One of the latter, representing a large dairy, told a reporter: "Our committee of seven, and Weimer was one of them, had the ordinance thoroughly overhauled, and we all conceded that what was left was quite satisfactory."[43] Weimer lost his case before the judge of the Municipal Court, who sustained the ordinance on the grounds that it enabled the health commissioner to keep a list of city vendors in the interest of the public health. Weimer attempted to carry his case to the Supreme Court, but his fellow milk dealers were too divided to proceed any further with the case. Although a few still clamored that they would "go clear to Washington if necessary for . . . their rights," most of the city dealers were ready to sign the registry and get on with their busi-

[40] *Common Council Proceedings*, July 2, 1888, p. 149; August 13, 1888, p. 201. See also *Sentinel*, July 10, 1888, p. 3; August 14, 1888.

[41] *Sentinel*, September 19, 1888, p. 3.

[42] *Sentinel*, October 3, 1888, p. 3; October 6, 1888, p. 3.

[43] Albert L. Vannaman, of the Hutchinson Milk establishment, quoted in the *Sentinel*, October 6, 1888, p. 3.

ness.[44] Finally Weimer himself paid his five-dollar fine and appeared at the health office to sign the register. By November 9, 1888 the *Sentinel* proclaimed, "The milk question in Milwaukee may be considered as settled."[45]

Milwaukeeans heaved a great sigh of relief, even though the price of milk jumped from five to seven cents a quart. Because of their anxiety about the potential control implied in registration, the dealers raised the price to help them to produce a product that might bear scrutiny by the milk inspector.[46] In fact the registry ordinance did not add significantly to health department powers, but Commissioner Martin, nearing the end of his term, did not raise the issue again. The milk dealers went about their business much as before, although now there were fewer of them. Many smaller dealers, unnecessarily worried about growing powers of the health department and fearing that possible added production costs would ruin them, closed their doors. The milk dealers listed in the city directory, only a small percentage of the total, decreased from 86 in 1885 to 45 in 1889.[47]

Milk consumers, more concerned about price changes than about quality, did not participate in the battles between the health department and the milk dealers. As one official from another city put it, the "public . . . is indifferent to its best interests and must be saved in spite of itself."[48] The Milwaukee health officers agreed.

During the 1880s the Milwaukee health department, for

---

[44] *Sentinel*, October 23, 1888, p. 3.

[45] *Sentinel*, November 9, 1888, p. 3.

[46] *Sentinel*, December 21, 1888, p. 2.

[47] *Wight's City Directory of Milwaukee*, 1885, pp. 35-36; 1889, p. 970. These figures indicate a small percentage of the actual dealers. By July 1889, 741 milk dealers had registered with the health department. *Sentinel*, July 26, 1889.

[48] Charles Harrington, Professor of Hygiene at the Harvard Medical School, in "The Practical Side of the Question of Milk Supply," *Pediatrics* 16 (1904): 209.

all its efforts, gained little more than the ability to identify milk producers in the city. When U.O.B. Wingate took over the health office in 1890, he immediately reopened the milk question, telling the common council that "as the laws are now, our inspections are of no practical use."[49] The health officer could not even tell which of the 904 registered milk vendors were still doing business in the city. Wingate, incorporating the ideas of Wight and Martin, wrote a comprehensive ordinance, which he proposed to the council as part of a package of reforms that he wanted to institute in the health department. This ordinance required that all city milk dealers be licensed, but only after the health commissioner approved the sanitary conditions of their cow stables. It disqualified from sale any milk produced by swill-fed cattle and required a solids content of nine percent, a considerable reduction from previous health department demands.[50] Introduced in June 1891, Wingate's ordinance breezed through its various readings and committees and passed into law in October. None of the political and economic battles earlier endemic to attempts to regulate milk broke out to mar the passage of Wingate's ordinance. The dairy association briefly revived, but did not actively resist, the measure.[51]

The easy passage of the 1891 milk ordinance may be explained in various ways. Milk dealers perhaps offered little resistance because they had come to realize that milk controls were relatively innocuous as practiced between 1889 and 1891. Also, there were fewer marginal dealers, those most vulnerable to the extra cost of reform and most active in earlier protests. The surviving dealers probably realized that the health department's strength and support in the city could no longer be successfully resisted. By 1891 the

[49] MHD, *Annual Report*, 1891, p. 31.

[50] *Sentinel*, October 20, 1891. See also, MHD, *Annual Report*, 1892, p. 42.

[51] The Milwaukee Milk Dealers Protective Association Articles of Association, #M556, October, 1891, Wisconsin State Historical Society.

health department had weathered much of the earlier indecision about its role, to win acceptance as a presence in the city. Increasingly, physicians and citizen organizations supported health department work. On a variety of issues, health commissioners had built up support, and Wingate depended on this basic acceptance to push for his reforms.

Wingate's success can also be attributed to the much-feared approach of cholera. This rapidly spreading, devastating disease caused municipal legislatures across the country to provide health expenditures and to allow expansions of power not permissible in ordinary times. Wingate took advantage of this climate of fear in 1891 and 1892 to win passage of legislation on hospitals, contagious diseases, and milk, all of which increased the power of the health department to control the health-related aspects of Milwaukee's life.

Wingate's impeccable reputation, his good relations with both the medical and political communities, and his skill in interpersonal relations generated personal support that helped the milk ordinance along. Unlike Wight and Martin before him and Kempster after him, all of whom offended at least some parties, Wingate alienated virtually no one. Although he did not get everything he wanted and found solutions to some problems elusive—for example, garbage—he managed to maintain an equilibrium and order that kept his communication channels open and made repeated successes possible.

Although Wingate achieved great success in passing milk legislation, he was unable to stop the injunctions that followed as some dealers fought the new controls in the courts. Wingate watched helplessly as deaths from cholera infantum and infantile diarrhea, diseases closely associated with milk, rose during the years of his administration.[52] (See Table 5-1.) When Walter Kempster took office in 1894, he, like his predecessors, was appalled at the foul state of the

[52] MHD, *Annual Report*, 1894, p. 7.

174

TABLE 5-1

Deaths and Percent of All Deaths from Cholera Infantum and Infantile
Diarrhea, Milwaukee, 1890-1893

	Cholera Infantum		Infantile Diarrhea	
Year	Deaths	%	Deaths	%
1890	121	3.23	250	2.24
1891	205	4.35	350	2.64
1892	213	4.71	345	2.49
1893	262	5.83	480	4.26

city stables and the adulterated milk sold in the city.[53] Upon
examining milk specimens, he found "particles of manure,
bits of cow's food, numerous hairs, mould, and fungus
growths, bits of insects, threads, human hair, moss and
other disgusting substances."[54] Finally, in 1896, the injunctions having been overridden, the council appropriated money for a milk inspector, and the health department
began effectively controlling milk sold in the city.[55]

The health department laboratory, created in 1896, developed as a very important element in milk regulation.
The inspector monitored production and distribution by
examining milk in the laboratory, trying to eliminate the
dirt and measuring the solid content to keep it high enough
to aid human growth. The milk inspector realized that the
health office also needed to promote refrigeration practices
to limit the growth of bacteria: "It is these germs which
decompose the milk and produce the most harm and cause
the most sickness, especially diarrhoeal diseases."[56] Despite
this understanding, the health department was not yet
equipped to undertake bacteriological studies, and until
1906 the food analyst relied on chemical tests to measure

[53] MHD, Annual Report, 1894-1895, p. 10.

[54] MHD, Annual Report, 1896, p. 10.

[55] Ibid., p. 68. See also Sentinel, June 19, 1896; June 20, 1896.

[56] MHD, Annual Report, 1896, p. 68. According to Shaftel, "A History
of Purification," the first time bacteria was mentioned by the New York
Health Department was also in the annual report of 1896.

fat and solid content and to determine the extent of watering or adulteration.

Even limited chemical analyses proved effective in regulating Milwaukee's milk supply. In 1896 the new laboratory found 204 samples of milk below standard for butterfat content. The health officer made seven arrests and imposed fines to achieve compliance with the laws. Kempster found that the fines had "a salutary effect," leading to "decided improvement in subsequent analyses."[57] When examining the morbidity and mortality statistics in the city, Kempster noticed that milk-related illnesses such as infantile diarrhea seemed most common in the fourteenth ward, where poor Polish immigrants predominated. He informally correlated this with his knowledge of the milk available in that ward—"milk which has had water little better than dilute sewage added to it"—and realized that the quality of milk differed around the city and that poverty seemed to exacerbate the risks.[58] Kempster could have bolstered his argument about the negative effects of bad milk by examining the city-wide age-related diarrheal deaths. The data for 1899 indicate no correlation between diarrheal deaths and deaths under the age of one year ($-.187$), supporting the supposition that most infants in Milwaukee were breast-fed. However, a very significant correlation connects diarrheal deaths with deaths under five years ($+.752$), suggesting that once children were weaned from the breast they became susceptible to harm from the contaminated milk supply.[59]

Kempster encountered many obstacles and frustrations in his attempts to "provide pure, wholesome milk to con-

[57] MHD, *Annual Report*, 1896, p. 69. For an account of the arrests, see, for example, *Sentinel*, July 13, 1896.

[58] MHD, *Annual Report*, 1896, p. 70.

[59] See Chapter 1 for further analysis of the patterns of disease in Milwaukee. The correlations were Pearson product moment correlations (r) and .752 tested significant at $>.01$.

sumers throughout the city."[60] Most people agreed that the health department should regulate the quality of milk, but there was disagreement about how much and under what conditions the department should determine milk production and distribution practices. The health officer still did not have free reign in deciding how to proceed, as Kempster discovered when he arrested several milk dealers for feeding cows distillery slop. He thought the 1891 ordinance permitted this, but the court, ruling that the "ordinance [was] unreasonable," freed the arrested dealers.[61] Kempster concluded that since it was "impossible to secure convictions," it would be best to outlaw cows from the city altogether.

Kempster realized the department was understaffed and could not effectively monitor the city's growing milk supply. He repeatedly petitioned the common council for additional staff and funds to regulate milk, but the council's generosity had stopped when it provided a milk inspector, who could be funded out of license fees. Kempster's lack of success with the council resulted at least in part from his previous conflict with that body over smallpox and garbage policy.

In addition to trying to enforce the 1891 ordinance, Kempster also sought to regulate how grocery stores marketed milk. He reported that in some shops: "large lumps of ice were found in the milk, to keep it cool—some of this ice is cut from ponds, and some from the Kinnickinnic— reeking with sewage filth, yet used thoughtlessly by shopkeepers to 'cool the milk,' the melting ice of course contaminating, diluting and poisoning the milk."[62] Kempster also wanted to include cream in the department's analyses. In 1898 he introduced an ordinance to increase the controls over cream and milk, this time including a new provision,

[60] MHD, *Annual Report*, 1896, p. 16.
[61] MHD, *Annual Report*, 1897, p. 12.
[62] *Ibid.*, p. 9.

Figure 20. Hokey-pokey ice cream vendor, Milwaukee, 1911. The Health Department had difficulty controlling production and sale of his unsanitary product.

tuberculin testing of cows. The council refused to pass it.[63]

Despite these frustrations, laboratory examination under analyst-physician W. S. Bennett progressed significantly during Kempster's tenure. Bennett reported that the fat content of milk showed continued improvements. By 1898 the department was analyzing 7,000 milk samples a year.[64] (See Table 5-2.)

Nonetheless, Kempster left office realizing that "the limit of life saving has not been reached in this community." He worried that unless the jurisdiction of the health department could be extended beyond the city limits, the milk supply could not be adequately monitored. The city examined milk samples as they crossed the city line for sale, but health department officers could not go beyond the city limits to inspect surrounding farms. Even some urban

[63] *Common Council Proceedings,* March 28, 1898, pp. 849-853.
[64] MHD, *Annual Report,* 1898, p. 108.

178

dairies still housed cows in ill-ventilated shacks, fed them distillery slops, and gave them no exercise. Opportunities abounded for persons with contagious diseases to handle and to contaminate the milk. Health department activity did not yet reflect the advances in medical understanding of disease transmission. For example, Kempster had proposed the required use of tuberculin tests to eliminate the threat of bovine tuberculosis, but the politicians, not yet convinced of the necessity of such expensive measures, relied on political and economic considerations in making their decisions. Kempster's parting message to the council in 1898, reflecting continuing tensions, put the blame for unfinished work squarely on the aldermen: "it remains for your honorable body to determine whether you will provide means to save these lives."[65] F. M. Schulz, after taking over the job of health commissioner, continued to despair of the milk situation. "So much needs to be done," he moaned, "that the outlook seems rather hopeless."[66]

TABLE 5-2[a]

Laboratory Work of Milwaukee Health Department, 1896-1899

	1896	1897	1898	1899
Number of analyses of milk made	2,134	4,675	6,919	6,318
Number samples below 3.0 percent	191	254	353	248
Percent below 3.0 percent	8.95	5.43	5.12	3.92
Average fat of milk	3.62	3.74	3.747	3.76
Number of analyses of cream made	37	54	102	276
Average fat of cream	19.20	19.40	17.80	20.95
Number of arrests made	7	138	148	106
Amount of fines imposed	$116.00	$810.00	$455.00	$830.00

[a] MHD, *Annual Report*, 1899, p. 95.

[65] *Ibid.*, pp. 17-18. As evidence that the jurisdiction question was of national concern see the discussion of the authority of cities outside city limits on the dairy inspection question in *JAMA* 34 (1900): 151-153. Topeka, Kansas, exercised authority outside its boundaries in the licensing of milk dealers, but Buffalo and Cincinnati had no jurisdiction except to refuse to allow contaminated milk to be sold in the city.

[66] MHD, *Annual Report*, 1900, p. 77.

The health commissioners' despondency and the imperfect condition of Milwaukee's milk spawned two private efforts in 1903 to upgrade the quality of city milk. The two exemplified the popular and conflicting theories about how to make milk safe: certification or pasteurization. Henry Coit, a New Jersey physician, founded and led the movement for certified milk.[67] Believing that clean raw milk provided the best nutriment for youngsters, he advocated spotless dairies, employees trained in cleanliness, and rapid, hygienic transportation to the consumer. Coit worked through physician groups to popularize the idea of certifying milk throughout the country. By 1912 there were 63 medical milk commissions with over 500 members. The strict monitoring necessary to produce certified milk made it difficult and expensive for cities to adopt the procedure on a wide-scale basis, but Dr. George Goler of Rochester, New York, proved through his municipal milk depots that it was possible to produce and to distribute certified milk to masses of people every day.[68]

For those who thought certified milk too impractical because of the expense and care of producing it, pasteurization proved an attractive alternative. This process merely

[67] For a thorough discussion of the certified milk movement, and Henry Coit's role in it, see M. J. Waserman, "Henry L. Coit and the Certified Milk Movement in the Development of Modern Pediatrics," *Bulletin of the History of Medicine* 46 (1972): 359-390.

[68] For standards set on certified milk, see Henry Coit, "The Code and Standards of the American Association of Medical Milk Commissioners," *Archives of Pediatrics* 29 (1912): 431-437. On Rochester, see George W. Goler, "Municipal Milk Work in Rochester," *Charities* 16 (1906): 483-487; "The Influence of the Municipal Milk Supply upon the Deaths of Young Children," *American Medicine* 6 (1903): 989-991; and " 'But a Thousand a Year' The Cost and Results in Rochester of Feeding Clean Milk as Food for the Hand-Fed Baby," *Charities* 14 (1905): 967-973; and Samuel Hopkins Adams, "Rochester's Pure Milk Campaign," *McClures Magazine* 29 (1907): 142-149. For more on certified milk see Martin Synott, "Public and Professional Confidence in Certified Milk and the Basis for this Confidence," *Archives of Pediatrics* 30 (1913): 774-778, and *Proceedings of the American Association of Medical Milk Commissions*, especially the years 1909, 1911 and 1915.

involved "cooking" or heating milk to kill potentially harmful bacteria. In 1897 Nathan Straus dramatically illustrated the beneficial effects of pasteurization when he instituted the heating process at the Randall Island Infant Asylum in New York: in one year the death rate fell by half.[69] The medical profession argued the merits of the two methods, but, because of the difficulties of monitoring clean raw milk, physicians increasingly advocated pasteurization.[70] In Milwaukee advocates of both certification and pasteurization tried to improve the city milk supply during the first decade of the twentieth century.

The Milwaukee Medical Society, responding both to the local situation and to the national efforts of Henry Coit on behalf of certified milk, established in 1903 a medical milk commission to supervise dairies and to provide clean raw milk for Milwaukee. Dr. Lorenzo Boorse, who spearheaded the society's activity, reported that "it was impossible at the present time to get a chemically and physiologically pure milk in the Milwaukee market."[71] In 1904 the commission opened a certified dairy in Pewaukee, a western suburb,

[69] The results were reported in most of the literature on milk in the early part of the twentieth century. See, for example, Spargo, *Common Sense*, pp. 230-233, or Parker, *City Milk Supply*, pp. 266-267.

[70] For discussions of the pasteurization processes, as well as comparisons, negative and positive, between it and certified milk, see, Rosenau, *The Milk Question*, Parker, *City Milk Supply*, Spargo, *Common Sense*, as well as numerous articles in the journals of the period: F. C. Gillen, "Pure Milk Supply and the Method of Obtaining It," *Transactions of Medical Society of Wisconsin* 30 (1896): 463-467; "Is Certified Milk Safe?" *JAMA* 58 (1912): 1200; Harns Moak, "Certified Milk v. Pasteurized Milk," *Proceedings of American Association of Medical Milk Commissions*, 1915. See also M. J. Rosenau, "Pasteurization," *JAMA* 49 (1907): 1093; Samuel Hopkins Adams, "The Solving of the Milk Problem: How Copenhagen has Established the Feasibility of a Pure and Healthful Supply," *McClures Magazine* 32 (1908): 220-227. Edwin O. Jordan, "The Municipal Regulation of the Milk Supply," *JAMA* 61 (1913): 2286-2291, reported a study conducted in 1910 by the AMA which reported pasteurization gaining in popularity in large cities. As much as 70 or 80 percent of the total supply was pasteurized in larger cities and it was gaining ground in small cities as well.

[71] Quoted in *Wisconsin Medical Journal* 1 (November 11, 1903).

giving the citizens of Milwaukee an alternative to city-supervised milk.[72] The farm produced certified milk under strict sanitary conditions: daily inspections, pure water, clean food, and good care for the tuberculin-tested cows, as well as frequent bacterial examinations and supervised, careful handling of the milk.[73] The milk commission charged fourteen cents a quart, a price only the upper classes could afford. But despite the limited distribution of its certified milk, the medical society felt proud of its contribution to "greatly decreasing infant mortality in Milwaukee."[74]

The second private effort in 1903 at improving Milwaukee's milk, the pasteurizing plant of the Children's Free Hospital Association, was also small in scale. The association distributed circulars printed in German, Polish, and English explaining how to secure and use pasteurized milk. Subsidized by Louis Gimbel, the Free Hospital charged only one penny for a pint bottle and provided rubber nipples for infant feeding. The hospital served as the major dis-

[72] The initial members of the commission included the following physicians: L. Boorse, H. V. Ogden, T. N. Hay, A. W. Myers, G. J. Kaumheimer, John M. Beffel, W. H. Neilson, E. W. Kellogg, and W. H. Bennett, representing the city health department. See Minutes of the Milwaukee Medical Society, November 25, 1902. For a copy of the resolution adopted by the Milk Commission, see Report of the Milk Commission, appended to the Milwaukee Medical Society Minutes for 1903. See also "The First Annual Report of the Milk Commission," appended to the 1904 Society Minutes.

[73] "First Annual Report of the Milk Commission," appended to the Milwaukee Medical Society Minutes of 1904. This report was presented to the society during its January 1905 annual meeting. The society had a little trouble with milk dealers trying to use the "certified" label to peddle regular milk, as letters in their files indicate. For example, in the correspondence books, see letter dated November 18, 1905 to Joseph G. Nuench of 22nd Street in the city who was labeling his milk wagons "certified milk," or another dated July 2, 1906. Both letters found in the Correspondence books in the Milwaukee Academy of Medicine archives.

[74] Milwaukee Medical Society Annual Report of the Secretary for the year 1905, read at the January, 1906, annual meeting. Appended to the 1905 minute book of the society.

tribution point, although settlement houses also participated by bringing pasteurized milk and nursing advice to the poorer Milwaukee citizens.[75] But, even with the aid of private initiative and a growing public constituency that favored milk reform, health commissioners experienced difficulty in achieving their goals. In 1898 the council had refused to pass Kempster's tuberculin testing regulation, and it continued to do so until G. A. Bading finally secured a tuberculin ordinance in 1908.[76] Immediately a group of dairy producers enjoined the city from enforcing the new law, calling the ordinance unconstitutional and discriminatory and maintaining that the tuberculin test itself was worthless. As their case, *Adams v. Milwaukee*, moved through the judicial system, each court, including the U.S. Supreme Court in 1913, sustained the ordinance.[77] Daniel Hoan, the future Socialist mayor of Milwaukee, who argued the city's defense, recognized that the "decision enlarge[d] the policy powers of the city to a great extent."[78] Not only was *Adams v. Milwaukee* important in establishing the right of the city to test cattle for tuberculosis; it gave those rights no matter where the cows lived, thus extending the powers of the health department outside the city limits. The issue that had hampered health controls over milk had been settled by the highest tribunal in the land, and the decision paved the way for increasing public health controls in the twentieth century.

In 1914 the new health commissioner, Dr. George C.

[75] *Wisconsin Medical Journal* 2 (1903): 183. See also *Sentinel*, July 26, 28, August 16, 19, 23, 1903.

[76] *Ordinances Regulating the Sale and Handling of Milk and Other Foods. Consisting of Portions of* "An ordinance to revise, consolidate and amend the General Ordinances of the City of Milwaukee Passed May 28, 1906," as amended by an ordinance passed March 30, 1908, p. 22.

[77] *Adams v. Milwaukee* (1913) 228 U.S. 572. The Wisconsin Circuit Court of Milwaukee County had handed down its decision on September 11, 1909, and the opinion of Referee Joseph G. Donnelly is reprinted in the MHD, *Annual Report*, 1909, pp. 163-188.

[78] Hoan was quoted in the Milwaukee *Journal*, May 12, 1913.

Figure 21. Bacteriology Laboratory of the City of Milwaukee Health Department, showing milk samples collected, c. 1910. Courtesy of City of Milwaukee Health Department.

Ruhland, who had been serving as assistant in the health office, attempted to enforce the now-vindicated tuberculin ordinance. In July he turned 6,000 gallons of milk back to farmers who had not complied with the law and who refused to promise to do so by the following summer.[79] This action resulted in a milk shortage when the milk producers, hurt by Ruhland's restraint, decided further to curtail Milwaukee's milk supply. In an attempt to break the health department authority they organized a boycott of the city. Ruhland tried in turn to break the dealer's strike by finding new suppliers.[80] While this confrontation progressed, the

[79] *Sentinel*, July 2, 1914.   [80] *Sentinel*, July 3, 1914.

hot weather aggravated the already difficult situation. Thirsty citizens increased their demands for milk at the same time the city was receiving only fifty percent of its normal requirements. South-side dealers, who usually used twenty-three wagons full daily, reported receiving no milk at all.[81]

Ruhland begged Milwaukeeans to tolerate the milk shortage long enough for him to force the city's usual suppliers to agree to the tuberculin test. But the milk dealers adamantly withstood his demands. Instead of supporting Ruhland, milk drinkers around the city showed decreasing tolerance for him and for Mayor Bading, who as health commissioner in 1908 had been responsible for the tuberculin ordinance. Before long the situation became desperate, and the aldermen, bowing to public opinion and the strength of the dairy producers, adopted a compromise position that simply postponed enforcement.[82]

One of the main issues in the 1914 "milk war" concerned pasteurization. Dealers who resisted the tuberculin test declared that pasteurization made the condition of the cow irrelevant because it destroyed the tubercle bacillus. Approximately eighty-five percent of Milwaukee's milk was being pasteurized by this time, and the dealers proposed a substitute ordinance making the heating process compulsory. Ruhland, working for acceptance of the tuberculin test, refused to support mandatory pasteurization as a substitute for tuberculin testing, even though he recognized the value of heating milk. Instead he put all his energies into getting the city to support his tactics and his position in favor of tuberculin.

Although the *Sentinel* supported the health department's attempts to force tuberculin testing on milk suppliers, the Socialist *Leader*, which usually favored expanding public controls, worked against the health commissioner. Believ-

[81] *Sentinel*, July 6, 1914.
[82] *Sentinel*, August 4, 1914; *Daily News*, August 25, 1914; *Common Council Proceedings*. See also *The Leader*, August 8, 1914.

185

ing that Ruhland wanted tuberculin testing only because of his loyalty to Bading, that the test was "antiquated," and that pasteurization would be more effective for conquering all diseases that could be transmitted by milk, the *Leader* actively supported the striking milk dealers.[83] The Socialists had their own complaints against Ruhland, who they believed had fired five sanitary inspectors merely because they were Socialists.[84] Out of office after enjoying two years of power in the city, Socialists attacked their opponents as "political doctors" and accused Ruhland and Bading of the basest motives when they allowed uninspected milk into the city in their attempts to break the milk dealers' boycott.[85]

Testimony before the common council revealed that city physicians supported both positions. Doctors agreed that the tuberculin test could be useful in detecting tuberculosis in cows and thus in preventing the transmittal of the disease from cows to humans. They were less clear about whether pasteurization prevented tuberculosis, but they agreed that it protected the milk supply from many other diseases.[86] Ruhland himself acknowledged that pasteurization could be beneficial. Nevertheless he maintained that it often did not kill the tubercle bacillus, because dealers, in their efforts to preserve the cream line that consumers valued, refused to heat the milk high enough or long enough. The health

[83] See *The Leader*, July, August 1914 passim.

[84] *The Leader*, June 27, July 21, 1914.

[85] See, for example, *The Leader* editorial, July 15, 1914: "If Bading and Ruhland and the political doctors have their way they will force the price of milk to 9 or 10 cents a quart without giving to the public any greater security than it has been getting from pasteurized milk for which it has been paying 7 cents a quart." This temporary break in the reform coalition resolved itself later in the milk controversy, but it illustrates the tensions that existed in the "progressive" alliance between groups that had major ideological differences.

[86] *Sentinel*, July 13, 17, 1914; *Leader*, July 11, 13, 14, 1914. The County Medical Society unanimously supported Ruhland's efforts to enforce the tuberculin test. For more reaction to the milk war see the "Milk" clippings box in the Milwaukee Public Library, Local History Room.

commissioner thus continued to reject the pasteurization ordinance as a substitute for tuberculin testing.

Ruhland's rejection of pasteurization confused the medical controversy for many Milwaukeeans. During the milk war the issues became dichotomized. Most people either supported Ruhland's efforts to compel compliance with the tuberculin ordinance or supported pasteurization as a substitute for tuberculin testing. In the hot August days when the city's thirst and tensions were high, the medical question about how best to protect the milk supply from carrying infection became political. Dealers, consumers, and newspapers, forced to take sides between tuberculin testing and pasteurization, most often opted for pasteurization because it promised protection against more than one disease. Ruhland garnered little support for his tuberculin enforcement battle, especially because people blamed it for causing the milk crisis.

The aldermanic compromise, forcing Ruhland to desist temporarily from enforcement, stilled the controversy, allowed milk to flow again at its normal volume, and prepared the way for Milwaukee's ultimate solution to the milk question. In 1916 the council passed a pasteurization ordinance (this time proposed by Ruhland), and after the First World War, which interrupted many city functions, enforcement of both tuberculin testing and pasteurization became possible. Previously competing forces, now freed from the political battlefield, united to support both measures to improve the quality of Milwaukee's milk.[87]

[87] MHD, *Annual Report*, 1916, pp. 147-148; *Pfeffer v. City of Milwaukee* (1920) 171 Wis 514. For a thorough discussion of the law and its relation to milk, see James A. Tobey, *Legal Aspects of Milk Sanitation*, revised edition (Washington: Milk Industry Foundation, 1947). For more on the health department reactions, 1914-1916, see MHD, *Annual Report*, 1914, pp. 16, 68-72; 78-81, 84; 1915, pp. 38-42, 48, 57-64; 1916, pp. 39-46, 67-71. See also the *Healthologist*, The Health Department Bulletin, October-November 1913, pp. 10-11; May-June 1914, pp. 2-5; September-October 1914, pp. 2-4; January-February 1915, pp. 9-10, 12; April 1915, pp. 13-16.

The milk war of 1914 convinced many health department officials that public education was more important
than political confrontation in paving the way for successful
public health measures. Many of Ruhland's troubles stemmed
from the failure of consumers to recognize the need for
tuberculin testing. "Users of milk," said one department
official, "have told us that they see no difference between
milk from a tuberculin tested herd and the milk from an
untested herd." Until consumers believed that such differences existed, they lacked the incentive to support health
reforms. The public would continue to support lower prices
and the traditional cream line until taught that unseen factors could be important in determining the quality of the
milk supply. This realization, continued the health officer,
"is compelling the Health Department to become an educational institution primarily, and is relegating the police
powers to a secondary place."[88]

The relationship between health department activity and
the morbidity or mortality of milk drinkers cannot be directly ascertained. Milk drinkers cannot be identified, except in a general way; varying enforcement levels cannot
be determined; milk dealers' compliance with attempted
reforms cannot be quantified. However, infant mortality
did drop during that period of the milk controversies, and
it seems likely that this decline was not unrelated to the
improving quality of Milwaukee's milk supply. It is sugges-
tive that the mortality of children ages one to five dropped
earlier than that for infants under age one. Contaminated
milk might have affected the age one-to-five group more
than it did the under-one group, who were still largely
breast-fed. The drop in the childhood mortality began in
the late 1870s, just when Dr. Wight began his crusade to
clean up Milwaukee's milk. Although he did not succeed
in getting Milwaukee's common council to pass any en-

[88] F. M. Luening quoted in J. Scott MacNutt, *The Modern Milk Problem*
(New York: Macmillan Company, 1917), p. 235.

forceable legislation, he nonetheless might have affected the quality of milk in the city. Dealers, fearing what might come, may have voluntarily begun policing themselves and producing cleaner and more healthful milk. Families, now aware of possible links between milk quality and child health, may have taken advantage of Dr. Wight's lists and evaluations of local dairies and stopped buying from the most offensive ones. After the passage of the 1887 and 1891 ordinances, all mortality under age five decreased, although the age one-to-five group again seems more responsive. Thus, although we cannot make precise quantitative correlations, we can conclude that, as intended, Milwaukee's campaign to improve the quality of city milk also improved children's health and life. (See Fig. 22.)

Figure 22. Infant Mortality as a Percent of Total Mortality, Milwaukee, 1870-1930, showing effects of milk legislation. *Sources*: Milwaukee Health Department Annual Reports and Division of Vital Statistics Records.

189

CHAPTER SIX

The Volunteers

During the nineteenth century, while the health department attacked the major public health problems of infectious diseases, poor sanitation, and tainted food, individual Milwaukeeans continued to face health crises of smaller proportions. When they fell sick, many residents turned not to the health department but to private associations, which grew in number toward the end of the century. Private philanthropy, by responding to the immediate needs of individuals, filled an important gap in municipal health services. With most physicians busy in their own practices and the health department concerned mainly with improving the environment and preventing epidemic disaster, those Milwaukeeans who could not afford the former and could not wait for the beneficial results of the latter relied on a vast array of private endeavors. In addition to providing free services for the sick, many agencies, either in conjunction with the health department or alone, emphasized preventive measures to keep the well healthy.

Some people from the educated upper half of the social strata tried to help the poor and uneducated, whose lives were burdened by physical hardships and bad health. Through an active club network, leisured women learned about the adverse social conditions in the city and initiated programs to eliminate some of the causes and effects of poverty. Settlement houses organized clubs, classes, and clinics to serve the growing needs of the urban poor.

Churches provided health and social services for their needy members. Civic-betterment clubs helped to bring order to the city, which was increasingly overwhelmed by expanding population and rapid industrialization. Medical societies, too, participated in private efforts to improve the public health, volunteering their members to serve on medical boards of numerous benevolent institutions.

Private organizations prospered in part because of the internal excitement they generated among their workers. The sight of scrawny, wan children filling out and growing rosy after a summer of good food and fresh air energized many volunteers to continue their labors. The voluntary associations also grew because their members, in addition to providing benefits for the poor, gained something themselves by their participation. German-Jewish residents in Milwaukee established the Hebrew Relief Society partly out of compassion for their central European brethren but also in the hope that the habits and appearance of the newcomers could be changed before they provoked an anti-semitic reaction.[1] Civic organizations supported urban cleanups and programs to improve citizen health, in part because an unhealthy city and unhealthy workers were bad for business, hampered city growth, and discouraged new enterprises. Many women found health reform an acceptable route to expanding their horizons beyond the home. Thus a combination of self-interest and altruism fueled most private participation in public health activities.

The early private agencies such as the Daughters of Charity, which opened Milwaukee's first hospital in 1848, worked largely alone. But as voluntary organizations proliferated toward the end of the nineteenth century, many health workers found that their efforts could be more effective if united with those of other agencies. Various Milwaukee

[1] Ann Shirley Waligorski, "Social Action and Women: The Experience of Lizzie Black Kander," unpublished M.A. thesis, University of Wisconsin, 1970, pp. 19-23.

charities and service organizations joined together under umbrella organizations such as Associated Charities and Central Council of Social Agencies to strengthen their ability to improve people's lives. But even these amalgamations proved unable to meet many of the threats to urban health. By the first decades of the twentieth century some voluntary agencies sought support from local government to bolster their programs and to provide central coordination for their efforts. Likewise, the city health department, realizing its inability to conquer Milwaukee's health problems without citizen support, sought out private organizations to contribute to public efforts. Thus a mutually advantageous relationship between the public and private sectors emerged in the early twentieth century, a relationship that adds an important dimension to our understanding of how American cities ultimately overcame some of the major obstacles to public health reform.

The most traditional charity organizations had religious affiliations. The Daughters of Charity, Milwaukee's oldest group to organize institutions for the sick, disabled, and needy poor, established St. John's Infirmary (later renamed St. Mary's Hospital) in 1848, to care for the sick "without distinction of class or religion or nation." Two years later, Daughters of Charity opened St. Rose's Orphanage for "destitute female children," most of whom in the founding year had lost their parents to cholera. During the ensuing years, these institutions expanded, and in 1877 the Daughters started St. Vincent's Infant Home, which provided shelter and care for needy mothers and infants.[2]

Although these Catholic sisters maintained the most substantial array of institutions for the destitute and sick in

[2] Peter Leo Johnson, *The Daughters of Charity in Milwaukee 1846-1946* (Milwaukee: Daughters of Charity, St. Mary's Hospital, 1946), pp. 38, 83-84, 141, 158, 197, 199. See also Philip Shoemaker and Mary Van Hulle Jones, "From Infirmaries to Intensive Care: Hospitals in Wisconsin," in Ronald L. Numbers and Judith Walzer Leavitt, eds., *Wisconsin Medicine: Historical Perspectives* (Madison: University of Wisconsin Press, 1981).

Milwaukee, many other agencies, such as Federated Jewish Charities, integrated health concerns into their activities.[3] Most representative of nineteenth-century charity work was Associated Charities, an agency that coordinated the work of many voluntary groups and carried out its own relief programs. Associated Charities preferred to work among those people who needed only minimal help. One of its favorite programs was a highly structured system of pre-paid and subsidized health care for people who could not otherwise afford medical services. This organization opened the Provident Dispensary in which "people of small incomes may, by the payment of small sums week by week, become entitled to the best attendance and all medicines when ill." Associated Charities also provided care for the "worthy" indigent, many of whom had become destitute because of sickness. In 1912, for example, 770 of the 1,349 aid recipients listed health-related causes of their "distress."[4] Case number 8406 (1890) offers an example of the kinds of people most frequently helped under the charity system:

"Last winter a poorly clad woman called at the office asking for aid, saying her husband was ill, that she could not leave him and go to work, but had to stay at home to care for him and was unable to earn anything. The case was investigated and found as reported.

"The case of this family was referred to the ladies of the Mission Band, who provided good medical attendance for the man and good food was furnished them. Gradually the man improved so his wife could leave him and go to work.

---

[3] See, for example, Federated Jewish Charities, *Annual Report*, 1905, pp. 18, 24, 32; and Charles Friend, "Tuberculosis Among the Jewish Dependent Poor of Milwaukee," *The Crusader*, January 1912, pp. 10-12.

[4] Associated Charities of Milwaukee, *Annual Report*, 1885, p. 12, 1912, p. 4. The 1912 causes of distress were: tuberculosis, 101; other illnesses, 375; physical disability, 80; insanity, 39; old age, 49; widowhood, 126; slack employment, 17; insufficient employment, 189; insufficient earnings, 43; desertion, 170; divorce, 10; shiftlessness, 50; bad management, 17; intemperance, 43; incompetency, 4; man imprisoned, 35; juvenile delinquency, 1; immorality, 1.

After a few months the man regained his health, a place was found where they both could work at good, fair wages and where they had a comfortable home. . . . This family has since laid aside money enough to buy a comfortable home and is now prosperous."[5]

The procedures common among private charities for scrutinizing applicants to identify the truly worthy limited the numbers and kinds of people that could be reached, but numerous beneficiaries testified to the value of the assistance.

As benevolent associations grew in number toward the end of the nineteenth century, their interests became increasingly specialized. Women's clubs in particular directed their reform efforts toward specific groups. The Woman's School Alliance, for example, concentrated on improving the conditions of the school buildings to enhance learning potential for students. Alliance members believed that they, as women, had a special contribution to make. "There are many features of school life which, while apparent to a mother's eye, are unnoticed by a father's," one wrote. The Alliance used this "mother's point of view" to improve ventilation and cleanliness and to eliminate "disease-breeding plumbing" in the schools. In the latter campaign the women worked closely with the health department and the medical society in a "war on the sewer gas."[6] Alliance members

---

[5] *Ibid.*, 1890, p. 11. See clipping from the Milwaukee *Journal*, November 15, 1942, on the development of the Mission Band in the WSHS pamphlet file on "Milwaukee Societies-Misc."

[6] Milwaukee *Journal*, February 22, 1895. As part of the miasmatic theory of disease, health officials in late-nineteenth-century Milwaukee believed that sewer gas—an invisible and odorless gas caused by the decomposition of organic material—could contaminate buildings connected to the city sewer system. According to Dr. Johnson, it could be "worse than a corpse in a house," transmitting to unsuspecting residents typhoid, diphtheria, and scarlet fever: "It gives no warning, and its unknown presence is not shunned. It sleeps with you, creeps into every cell of your lungs, and lays shadowy fingers on every drop of your heart's blood." MHD, *Annual Report*, 1878, p. 85. Ridding the city of this danger became an important part of health department work.

responded favorably when Health Commissioner Kempster, acknowledging that "women when they put their mind to it can accomplish anything," sought their aid in getting regular school medical inspection.[7]

The Milwaukee Children's Hospital Association, "devoted solely to the relief of sick children," opened a free hospital in 1894 to serve "all patients under 15 years of age and not afflicted with uncurable or very infectious diseases." This women's group concentrated on creating a "home life" in the hospital, where "children are guided by a firm yet gentle hand."[8] The Women's Fortnightly Club created and staffed the Babies Fresh Air Pavilion. Their tents on the banks of Lake Michigan allowed poor women and children to escape from the sweltering heat of ill-ventilated tenements and provided fresh milk and nursing care to the ailing.[9] Still another group of women, most of them wives of physicians, organized the Mothers' Cottage Association to build small cottages on the city hospital grounds so that mothers could stay with their children when the young ones contracted contagious diseases.[10]

The Milwaukee Maternity Hospital Association, organized in 1906, provided "proper medical care and nursing for poor women during confinement," as well as general hospital and dispensary facilities for women and children. By 1911 its medical and nursing staff annually attended 191 births (half of them at home and half in their hospital) and treated 1,578 women and children at their dispensaries. Jewish and Catholic women took greatest advantage of the birth services. Emphasis on prevention and education led the Maternity Hospital Association to cooperate

[7] *Sentinel*, February 26, 1898.

[8] *Journal*, February 22, 1895. See also Milwaukee Children's Hospital Association, *Annual Report*, 1910-1911 (WSHS) which gives the medical report, diagnosis, and disposition of 659 people treated that year.

[9] MHD, *Annual Report*, 1914, pp. 18-19.

[10] *Sentinel*, August 24, 1892, p. 3; September 28, 1892. The women included "Mrs. Doctors" H. M. Brown, A. B. Farnham, William T. Batchelor, and U.O.B. Wingate.

Figure 23. Babies Fresh Air Camp on the lake front, c. 1913. Courtesy of City of Milwaukee Health Department.

with other philanthropic groups and with the health department to serve the community health needs. The members participated in child welfare work in close connection with the Child Welfare Commission and, to train the next generation in hygiene and infant care, originated "little mothers classes," which the city eventually took over.[11]

Milwaukee's settlement houses offered numerous health services for the city poor. The University Settlement, founded in 1902, served the crowded south side with activities designed to Americanize and to advance the resident Polish immigrants. The settlement initiated a visiting nurse service

[11] Milwaukee Maternity Hospital and Free Dispensary, *Annual Report*, 1911, 1917, in the pamphlet file of the WSHS. A general description and assessment of Milwaukee's women's club's activities can be found in Harriet Noel Fritsche, "A Comparison of the Activities of Women's Organizations in Milwaukee from 1890-1895 and 1910-1915," Bachelor of Philosophy thesis, University of Wisconsin, 1928.

in the neighborhood, which, in the tradition of "trying to put itself out of business by pushing over on some other agency everything it could," it transferred to the Visiting Nurse Association. The settlement also initiated a program for crippled and sick schoolchildren that it subsequently passed on to the school board.[12]

Over fifty volunteers staffed the University Settlement, teaching neighborhood children and parents, providing library facilities, organizing clubs, and giving advice about baby care and child rearing. Winnifred Salisbury, who worked at the settlement in 1906, surveyed housing conditions in the fourteenth ward and found "over-crowded quarters that are always too hot, too cold, or too damp." She took a special interest in the factory work of young women. To investigate conditions firsthand, she worked in a candy factory. In her report to the University Settlement, she described workers who washed the filthy work tables only once a year, held caramel against their greasy clothing, and allowed machine oil to drip into the candy vats. "One girl had a pet habit of holding a 'sucker' in her mouth for a minute or two before wrapping it in its clear, white paper, and tossing it into its box"; another worker cooled her flushed face with the papers before wrapping the sweets. Salisbury found the ten-hour workday and the stifling environment so physically demanding that she fainted before she had been on the job a week.[13] Her study of factory conditions presaged the later health department program of factory inspections.

The Abraham Lincoln Settlement, which grew out of the Ladies Relief Sewing Society and the Milwaukee Jewish Mission and Night School, played an equally important role

[12] University Settlement, "A Review" (1908) and "After Ten Years" (1912) in the pamphlet file at the WSHS.

[13] Winnifred Salisbury, "Some Work Done at the University of Wisconsin Settlement," July, 1906, in DeWitt Clinton Salisbury Papers, 1972 additions: Winnifred Salisbury correspondence and miscellaneous, 1890-1958. WSHS.

in promoting health programs. It opened its doors in 1900 under the energetic leadership of Lizzie Black Kander, and, like the University Settlement, organized classes for adults and children, club activities, and a circulating library. One of the settlement's most popular attractions was its public bath, which, along with the *Settlement Cook Book*, provided the funding for the rest of the settlement programs. The bath was probably the most popular of the efforts to maintain health in the city environment. To its patrons, it could be as enjoyable as the proverbial local swimming hole. To its promoters, the bath provided necessary soap and water, unavailable elsewhere, to help people to ward off the menace of filth and disease. Kander tried to educate and to uplift the settlement's poor neighbors and to coordinate numerous private efforts to foster public health and welfare. She appeared at various women's club meetings to seek help for city-maintained public baths, actively supported the city child welfare stations, and advocated open-air public schools for tuberculosis relief and prevention.[14]

Kander and other settlement workers in Milwaukee hoped to add a human element to public efforts to promote health at the same time as they carried on their own independent functions. They incorporated public baths, milk stations, and tuberculosis clinics into their crowded quarters because, through their daily contact with the poor, settlement workers understood the necessity of such programs. Settlement activities depended on a local assessment of neighborhood needs; they changed as needs changed or as other agencies took over particular services. Settlement workers provided a personal approach to individual problems, and the people who shied away from governmental interference in their lives tolerated and even welcomed neighbor-

[14] Lizzie Black Kander papers, WSHS, Box 1; Federated Jewish Charities, *Annual Report*, 1915, pp. 88, 90; Milwaukee Social Science club, Minutes, December 5, 1911, WSHS. On the public baths, see M. T. Williams, "The Municipal Bath Movement in the United States, 1890-1915," unpublished Ph.D. dissertation, New York University, 1972.

hood-based efforts. Settlements provided excellent locations for city health projects and coordinated the work of
various groups within their local setting.[15]

The Visiting Nurse Association, incorporated in 1907 by
Sarah Boyd, one year after she had hired a Chicago nurse
to live in her home and to care for her "less fortunate
neighbors," was from its inception intimately involved in
public health work in Milwaukee. The nursing staff, financially sponsored by prosperous Milwaukeeans, quickly expanded and pledged itself to "give skilled nursing care and
health instruction to the sick in their homes regardless of
race, color, creed, or ability to pay."[16] The nurses cared for
tubercular patients, examined children in the public schools,
and ran a summer day camp for sick babies, responding
to calls from doctors, charities, hospitals, settlement houses,
ex-patients, private individuals, and the health department.
By 1911 fourteen nurses annually made 16,871 calls and
cared for 2,625 patients. The growing staff of trained women
became an important link in the health care services available to Milwaukeeans who could not afford private care.
A major emphasis was preventive. "[T]here is no [need]
greater than the welfare of the little babies," said the superintendent of the visiting nurses in 1910. "We want to

[15] See Allan F. Davis, *Spearheads for Reform: The Social Settlements and the
Progressive Movement 1890-1914* (New York: Oxford University Press, 1967),
who makes a case for reformist ideology among settlement workers, and
Ann Shirley Waligorski, "Social Action and Women," who accepts the ad-
hoc reformist position to explain settlement programs. See also Katharine
Bement Davis, "Civic Efforts of Social Settlements," *Proceedings of the National Conference of Charities and Correction* (Boston: Geo. H. Ellis, 1896),
pp. 131-137; James B. Reynolds, "The Settlement and Municipal Reform,"
ibid., pp. 138-142; Charles C. Cooper, "The Service that Settlements and
Neighborhood Houses may Render in the Community's Plan of Child
Protection," *ibid.*, 1915, pp. 193-197.

[16] *Milwaukee Visiting Nurse Association 1907-1927*, pp. 9, 11, in the Papers
of the Visiting Nurse Association of Milwaukee, WSHS, Box 1. See also
"Articles of Association" (June 7, 1907), *Annual Report*, 1908, p. 3, Reel 1
of the Papers of the Visiting Nurse Association of Milwaukee, WSHS.

reach the babies before they are ill, teaching the mother to care for her child and avoid the preventable diseases which claim so many." To best accomplish this goal, the Association affiliated with the Child Welfare Commission in 1911 (see Chapter 7) and contributed the services of two nurses to that body.[17] Other association services, such as the summer camps for sick babies, similarly became integrated into the health department system.[18]

The Milwaukee Medical Society, established in 1845 for the social and professional advancement of its members, also involved itself in community health reforms. The medical society members staffed numerous private facilities for the poor, such as the Maternity Hospital and the Children's Free Hospital. Physicians promoted major public health endeavors, including the Medical Milk Commission and school medical inspections. The health commissioners who were society members brought their problems and projects to their colleagues to solicit help. City programs for hospitals, dispensaries, vaccinations, and school health all received support from organized physicians, who had formed the original boards of health in the 1840s and who maintained their commitment to public programs.

Only rarely did Milwaukee Medical Society members oppose public health efforts. Answering one health department demand for reporting tuberculosis in 1906, Dr. Horace Manchester Brown indicated why private practitioners could not always support public programs. He refused to report tuberculosis to authorities since he would not be paid to do so. More important, he argued that such demands were unfair to private physicians: "Suppose we see a young woman with enlarged glands of the neck; we report the case as one of tuberculosis; the man to whom she is

[17] "Superintendent's Report," 1910, following page 38 in the Record Book of the Visiting Nurse Association, Reel 1. See also the *Annual Reports*, 1907-1930 and "Twenty Years of Service 1907-1927," WSHS.
[18] Visiting Nurse Association, *Annual Report*, 1911, following page 46 in the Record Book of the Visiting Nurse Association, Reel 1, WSHS.

engaged hears of it and breaks the engagement. Do we not lay ourselves open to action for damages? And if we do not report it the Health authorities will try to fine us. The medical man is pounded on both sides."[19]

Similarly, society members argued against school advice stations in the 1920s because of their potential intrusion into private practice. Although public health efforts occasionally collided with the private interests of physicians, physicians as a group and as individuals usually supported the city physicians who worked for the public's health.

The Medical Society of Milwaukee County entered into public health work after a particularly provocative 1904 symposium emphasized the preventability of tuberculosis. After the presentations Dr. Holt E. Dearholt fired a challenge to his colleagues: "Well, gentlemen, what are you going to do about it?" The society appointed a tuberculosis commission to study the problem and to propose solutions. Many of its members had a personal as well as a professional interest in tuberculosis: Dr. John W. Coon had lost his sweetheart to the disease; Dr. John M. Beffel had been orphaned by it; Dr. C. H. Stoddard had lost his brother to it; Dr. Gilbert Seaman's two brothers had died of it.[20] The commission's first effort was to arrange for a twelve-day show in downtown Milwaukee for the national tuberculosis exhibit, which dramatically illustrated for an estimated 53,000 people that tuberculosis was communicable, preventable, and curable.[21] "I can still see John Beffel leading us from exhibit to exhibit," recalled one man of his schoolboy visit.

[19] Milwaukee Medical Society, Minutes, October 23, 1906. Milwaukee Academy of Medicine.

[20] Louise Fenton Brand, *Epic Fight: Wisconsin's Winning War on Tuberculosis*, unpublished typescript, WSHS archives, pp. 77-78; *Wisconsin Medical Journal*, May, 1904, pp. 736, 751. The personal interest of the commission members is noted in Harold Holand, *House of Open Doors* (Milwaukee: Wisconsin Anti-Tuberculosis Association, 1958), p. 27.

[21] Brand, *Epic Fight*, pp. 128-145; Hoyt Dearholt, "History of the Movement Against Tuberculosis in Wisconsin," *The Crusader*, October 1910, pp. 3-7.

"How those eyes could flash! How that voice could ring! Most of us had gone to that exhibit expecting to be bored. We came away in a far different mood."[22]

The energy that the physicians put into the education exhibit also permeated their efforts in 1907 to build a local tuberculosis sanatorium on the western outskirts of Milwaukee. Their continuing campaign led them to join with other groups and to launch the state-wide Wisconsin Anti-Tuberculosis Association in 1908. The association's monthly journal, *The Crusader*, maintained communication among the various groups around the state; newspapers distributed its message widely; pamphlets and posters plastered public spaces; and association leaders kept up a busy lecture schedule as public education became the fledgling Wisconsin Anti-Tuberculosis Association's primary goal. The association carried out a survey of cases in the city and helped to pressure the city government to appoint a Commission on Tuberculosis, which eventually became the tuberculosis division of the city health department.[23]

The enormous proliferation of private agencies and organizations attempting to alleviate the adverse social and physical conditions of the turn-of-the-century city prompted the creation of an umbrella organization, the Central Council of Social Agencies, a "federation of social philanthropic and civic organizations united to further cooperation among agencies working for community welfare, to prevent duplication of effort, to study social problems, to promote

[22] Theodore J. Werle, who later served on the staff of the Wisconsin Anti-Tuberculosis Association and then moved to a similar position in Michigan, quoted in Brand, *Epic Fight*, p. 136.

[23] Kathrene Gedney, "The Survey of Milwaukee," *The Crusader*, July 1910, pp. 1-3. See also Holand, *House*, pp. 59-61; Edith Shatto, "City Government Fighting Tuberculosis," *The Crusader*, July 1911, p. 12; Edith Shatto, "The Crusade in Milwaukee," *The Crusader*, October 1911, pp. 26-30; and Sarah West Ryder, "Safeguarding City Health: The Growing Significance of the Health Department," *The Crusader*, October 1911, pp. 31-32. See also Brand, *Epic Fight*, pp. 235-238.

social reforms and to assist in activities for the general welfare." The council, like its constituent members, was active in health reform. In 1912, for example, its public health committee helped to increase the number of city health inspectors from fifteen to twenty-two, aided in the repairs of the old Isolation Hospital, supported the County Tuberculosis Hospital, broke up a notoriously bad baby farm, and endorsed legislation on regulating nursing practice and sputum disposal. The following year the committee began a campaign to introduce social services into the city's hospitals, worked in conjunction with the Milwaukee Medical Society and the City Club for a new general hospital, secured a psychopathic clinic at juvenile court and a venereal disease clinic at Marquette University, and assisted the medical societies in framing an amendment to protect the insane and feeble-minded. This agency coordinated efforts with the health department by encouraging the health commissioner, school physician, and child welfare director to serve as ex-officio members on its health committee.[24]

Even the Central Council, however, could not overcome the limitations of private benevolence, which continued to address the immediate needs of relatively select groups of

[24] Central Council of Social Agencies of Milwaukee, *Articles of Incorporation and By-Laws*, n.d. in WSHS pamphlet file. Among the sixty constituent societies, the following carried on specifically health-related work: Associated Charities, Child Welfare and Charity Committee of the Daughters of the American Revolution, Children's Home Society of Wisconsin, City Club, Council of Jewish Women, Federated Jewish Charities, Fortnightly Club, Ladies Sanitary and Benevolent Society, Marquette Dispensary Clinic, Milwaukee Children's Hospital, Milwaukee County Medical Society, Milwaukee Infant's Hospital, Milwaukee Maternity Hospital, Milwaukee Society for the Care of the Sick, Milwaukee Visiting Nurse Association, Mission Band, Settlement Association, Social Economics Club, University Settlement, Wisconsin Anti-Tuberculosis Association, Wisconsin Blind Association, Woman's School Alliance. See also Central Council of Social Agencies, "A Review" (1915) in the Marion G. Ogden Papers, WSHS, Box 3, folder 5. Marion Ogden was a Milwaukee social worker and activist who served for years with the Children's Betterment League and the Milwaukee Boy's Club.

people. Although a large number of associations were active in Milwaukee, a small number of volunteer leaders did most of the work.[25] The scarcity of doctors able to donate time to private philanthropic associations limited the pool of available experts. Groups trying to curb tuberculosis had no choice but to ignore the cases of malnutrition or diphtheria they encountered. School health workers bypassed aspects of diseases fostered at home. The jurisdiction of settlements remained limited to immediate geographic boundaries. Insufficient finances circumscribed the visiting nurses' work. While the coordinating council could alleviate some of the overlap of functions, it did not have the authority or the money to overcome the financial and personnel limitations intrinsic to the decentralized private endeavors. A city-wide private program adequate to meet the health needs of the city was not feasible.[26]

In the early twentieth century, voluntary organizations and the city health department moved toward cooperation because both could benefit from working together. The health department needed wide-scale political support for its activities, and the voluntary associations needed money and centralized coordination to make their efforts worthwhile. Increasingly, private associations tried to get their projects—visiting nurses, little mother's classes, public baths, neighborhood clinics, and summer outdoor programs—integrated into the city's public health program. Such connections freed the private agencies to expand their programs in new directions and allowed the health department

[25] Interlocking directories characterized Milwaukee's benevolent associations. See their papers in the WSHS. Lizzie Black Kander was associated with at least eight women's and religious organizations and Marion Ogden was linked to 13 or more service organizations.

[26] I reached the conclusions about the limitations and overlapping functions of private philanthropy from the City Club 1919 investigation of school medical inspections, City Club Papers, Box 3, WSHS. See also Stephen Smith, "Uses and Abuses of Medical Charities," *Proceedings of the National Conference of Charities and Correction* 1898 (Boston: Geo. H. Ellis, 1899), pp. 320-327.

to broaden and centralize services.[27] Especially after 1910 city health officials, realizing that cooperation rather than confrontation won public support for their programs, created channels through which public and private activity could be unified in a single quest for improving citizen health.

The health department in the Socialist years 1910 to 1912 worked particularly hard to link the work of private relief agencies with its own and to provide a centralized focus for Milwaukee's health efforts. Realizing that "private philanthropy has been wrestling" with health problems "for many years," city treasurer Carl D. Thompson observed that what was needed was "some central organization to coordinate these various forces and to supplement them with such assistance as will make their work more effective."[28] Health Commissioner F. A. Kraft worked precisely for these goals. He did not seek to take over the efforts of private

---

[27] This discussion has been limited to the activities of a few groups, but it is not meant to suggest that they were the only ones interested in public health. The Woman's Club of Wisconsin, devoted to "intellectual culture," listened to many lectures by their own Dr. Helen Bingham about public hygiene and also invited Walter Kempster, C. N. Hewitt, and other notable public-health-oriented physicians to speak before them. The Beta Study Club, the Milwaukee College Endowment Association, and the Milwaukee Social Science Club likewise heard lectures on the subjects of women in public health work and the importance of clean city environments. (Woman's Club of Wisconsin, *Annual Reports*, 1887-1923. See especially 1916-1917, a special report on the club's fortieth anniversary; Beta Study Club, 1897-98 Report, Milwaukee College Endowment Association, 1896-97 Report; WSHS.) The Milwaukee Society for the Care of the Sick operated four part-time dispensaries around the city for patients who could not pay for medical services. For more on Milwaukee's charities, see William George Bruce, *History of Milwaukee City and County* (Chicago: S. J. Clarke Publishing Co., 1922), Vol I, pp. 739-753; and the *Directory of Social Welfare Organizations: A Handbook for Citizens* (Milwaukee, 1912, 1920, 1924).

[28] Carl D. Thompson, "Health Care in Milwaukee," *Coming Nation*, August 19, 1911. I would like to thank Paul Buhle for calling my attention to the series of articles Thompson wrote on Milwaukee health. See also August 5 and 12, 1911.

agencies, but merely to fit them into a broader framework, with coordinated responsibility where he thought it belonged—in the health department.

The activities of the Socialist-appointed health officials fostered greater community participation. Of "primary importance" to Kraft was public education. "It is not possible to achieve proper and lasting results," said Dr. Kraft, "save by winning public cooperation."[29] Kraft and his staff initiated newspaper advice columns, distributed health care pamphlets, and produced a monthly bulletin aimed at enlightening the general public about health protection and disease prevention. The health department encouraged new programs that fostered close bonds with ongoing community agencies. The Child Welfare Commission consolidated the work of numerous private organizations within the health department in 1912. (See Chapter 7.) This relieved the Visiting Nurse Association of the financial burden of carrying two nurses and freed it to expand into school and factory work.[30] The tuberculosis division of the health department, begun in 1912, helped to lighten the obligations of the Wisconsin Anti-Tuberculosis Association and marked the beginning of the city's "official recognition of its responsibility in dealing with tuberculosis as a community problem."[31] The city division incorporated the work of the Society for the Care of the Sick, which ran tuberculosis dispensaries, and the Visiting Nurse Association, which staffed those dispensaries, integrating them in the larger city-wide attack on disease. Many of the city's health programs won public support precisely because they freed private associations from some financial burdens and because the integration of efforts increased both public and private effectiveness.

As the bonds between the public health department and

[29] MHD, *Annual Report*, 1911, p. 13.

[30] Visiting Nurse Association, Report, November 1, 1914, following page 76 of Reel 1 of their Papers, WSHS.

[31] Brand, *Epic Fight*, p. 235.

various private organizations grew, health achievements flourished in Milwaukee. A sanitary inspector who had worked in the health department under three previous health commissioners concluded in 1911: "There were times . . . when there was more done that raised a disturbance and got headline notices in the papers, but never as much quiet, consistent, effective work" as was accomplished under Kraft.[32] Although the Socialists were defeated in the municipal elections of 1912, the consolidation of the efforts of private organizations—such as the Wisconsin Anti-Tuberculosis Association, the Visiting Nurse Association, and the settlements—into municipal programs continued well beyond that year. Few Milwaukeeans accepted Socialist Thompson's analysis that public health was a "labor problem" that necessitated major alterations in the capitalist economy, but most Milwaukee residents agreed with him that effective public health work depended upon the cooperation and understanding of a broad coalition of citizens.

The kinds of private-public interactions that led to successful public health reforms in Milwaukee can best be understood by examining the communications between the city health department and the public health committee of the City Club of Milwaukee. An outgrowth of the local chapter of the National Municipal League, the City Club was a nonpartisan reform organization devoted to civic improvement. Its leaders represented the diverse elements of Milwaukee's reform movement: Progressive Republicans, liberal Democrats, Socialists, and Social Gospel Christians. Physicians played an important role in the formation of

---

[32] Joseph Derfus, speaking with Carl Thompson in *Coming Nation*, August 19, 1911. For national sentiment along the same lines, see James F. Crichton, "The Co-ordination of Official and Private Health Agencies," *Proceedings of the National Conference of Charities and Correction*, 1913 (Fort Wayne, Indiana: Fort Wayne Printing Company, 1913), pp. 165-168; John A. Kingsbury, "Coordination of Official and Private Activity in Public Health Work," *ibid.*, pp. 169-173.

the club in 1908, and eighty-five of them belonged by 1914. The club's first secretary was Dr. Fernando Mock, who wrote health columns in the Milwaukee *Journal*. From its earliest days the City Club endorsed or actively worked for city sanitation and public health as part of its campaign to improve city life. The club sought school medical inspections, open-air schools, housing improvements, a new sewage treatment plant, and improvements in garbage collection. In 1916 it sponsored a sickness survey to reveal the gaps in health care in the city.[33] The methods by which the club carried out these activities disclosed how extensively the private and public sectors interacted in the process of achieving health reform in the early twentieth century.

The Milwaukee Medical Society helped to start the first school medical inspections by convincing the school board of the program's value in 1909. By 1911 the health department had begun a concurrent program of inspecting parochial schoolchildren. In 1916 Health Commissioner Ruhland, wanting to expand and to systematize the health inspection of parochial schoolchildren, wrote to the secretary of the City Club, Hornell Hart: "For obvious reasons, I would like, if possible, to obtain the endorsement of the City Club for the plan. . . . May I depend upon you to present this matter for me and urge its endorsement?"[34] After consulting with the Committee on Public Health, the civic secretary replied that it was ready to take "a stand favorable to the establishment of such inspection." Ruhland then pressed his point even further: "I should like to know if I may depend upon you . . . to appear before the Council Committee when this matter is brought before them."[35] After the successful council debate, in which the City Club

[33] A comprehensive analysis of City Club early activity can be found in Roger Roy Keeran, "Milwaukee Reformers in the Progressive Era: The City Club of Milwaukee 1908-1922," M.A. thesis, University of Wisconsin, 1969. See especially pp. 69-71; 76-77 for health-related activities.

[34] City Club of Milwaukee Papers, Box 3, WSHS.

[35] *Ibid.*, letter of November 9, 1916.

participated, Ruhland wrote Hart again: "I wish to thank you, and thru you, the City Club, for the valuable assistance you have given in the matter of establishing medical inspection in parochial schools."[36]

Ruhland and Hart established a pattern of communication that effectively preempted much of the open controversy that had characterized the introduction of health reforms in the nineteenth century. Instead of airing possible differences in the press or attempting to quash any opposition that might arise, the health commissioner now summoned help from influential friends and built his case before it surfaced publicly. The City Club, too, was pleased to be consulted and to have influence over the form and content of proposed changes in the municipal government's activities.

Letters between the City Club and the health department reveal that health programs could originate from either side and that the health commissioner came to rely heavily on City Club investigations and ideas for program development. Ruhland wrote, for example, seeking "suggestions for a health program or whatsoever health work" the City Club thought advisable for the department to undertake.[37] When the health department proposed to unify the school medical inspections under its own jurisdiction in 1919, Ruhland again wrote the civic secretary, now Leo Tiefenthaler, to request "the advantage of an additional investigation" into the contemplated changes.[38] The City Club response came after a thorough investigation of the problem that took almost three months and that produced a sixteen-page printed pamphlet analyzing the problem of "Medical Inspection in the Schools of Milwaukee."[39]

[36] *Ibid.*, letter of December 5, 1916.
[37] *Ibid.*, Box 8, letter of April 3, 1919.
[38] *Ibid.*, letter of April 8, 1919.
[39] The City Club of Milwaukee Committee on Public Health, "Medical Inspection in the Schools of Milwaukee: A Study" (Milwaukee: September 1919). The Report was produced by a sub-committee composed of Eliz-

The City Club's major concern was "in securing a larger and more effective program for medical school inspection in the schools of Milwaukee and only secondarily in the question of what department of the city government, the Health Department or the Board of School Directors shall have charge of the work."[40] In the process of recommending to the common council an expanded corps of doctors and nurses to examine schoolchildren and of suggesting that physical examinations be made "upon the bare chest of the child" to ensure a useful diagnosis, the investigators concluded that lodging the inspections in two departments created expensive duplication of services. Because they believed school health was "only one phase of the general community health problem," the investigators recommended that the consolidated program "be placed in the hands of the Health Department."[41] Ruhland was delighted with the comprehensiveness of the City Club study and even more so with its conclusions. "I appreciate particularly the recommendations made by the committee," he wrote to Tiefenthaler, "inasmuch as they coincide with plans which this department has in mind for the enlargement of its service. I trust," he continued, "that the committee may come to the support of the department at the time of budget making."[42]

In addition to its interest in school health, the City Club supported a sewage treatment plant to relieve the pollution of the Milwaukee harbor and to improve the quality and healthfulness of city water. When the "question of the obnoxious taste in the water" became critical in 1918 and 1919, the club brought representatives of the companies most responsible for water pollution together with the city

---

abeth G. Upham, and Drs. H. E. Dearholt and Samuel Higgins, with the help of public health instructors from the Wisconsin Anti-Tuberculosis Association.

[40] *Ibid.*, p. 2.
[41] *Ibid.*, pp. 10, 12.
[42] City Club Papers, Box 8, letter of July 23, 1919.

and state representatives to solve what was becoming a serious menace to the public health. The City Club secretary
participated in a series of meetings among the "offending
companies," the aldermen, the city and state health officers,
and the mayor, and helped them to reach a compromise
to shut down some plants and improve conditions in others.[43]

One of the most innovative projects of the City Club
Public Health Committee was the planning and execution
of a city-wide health survey. Sampling all parts of the city—
rich and poor, native and foreign—the club secured data
on the condition of Milwaukee residents on October 26,
1916: How many were sick? Did they have medical care?
Were they wage earners? What was the nature and length
of the sickness? Were there any sick benefits? In order to
obtain answers to these questions, committee members organized over one hundred volunteers from fourteen municipal and private agencies, gave them a dinner and training session, and sent them out to take a census "in
representative blocks scattered throughout the city."[44] The
social workers, nurses, and students who knocked on the
doors of homes throughout the city carried letters of support from the mayor and the archbishop, translated into
languages appropriate for their sections of the city. The
newspapers advertised the event and "cooperated very cor-
dially" in promoting its "splendid success."[45] The survey
reached a sample of 2,500 families and showed that ten
percent of the people in Milwaukee were sick on October
26, 1916, that less than half of those sick were under a
doctor's care, and that the poor were "seriously sick three

[43] See Leo Tiefenthaler's notes about Waterworks and Sewage in the
City Club Papers, Box 3, November 1918 to April 1919, and the Reports
on Sewage Disposal prepared for the City Club, n.d. See also the City
Club *Bulletin*, 1916, passim.

[44] City Club Papers, Box 3, "Public Health Correspondence" folder,
"copy" for article to appear in the City Club *Bulletin*, Vol. 2, October
1916.

[45] *Ibid.*, "copy" for the City Club *Bulletin*, Vol. 2, November 1916.

times as frequently as the well-to-do." The City Club data analysts considered the sickness figure low since people reluctantly reported venereal diseases or tuberculosis, both of which carried a social stigma. They calculated a loss in wages of approximately $3,000,000 per year for the city.[46]

The City Club, either on its own initiative or at the request of the health commissioner, advocated open-air schools as a "means of building up health and vigor to resist tuberculosis," supported budget increases for health department work, and investigated conditions at the city hospital and recommended renovations so "it may serve the reasonable requirements of a city of almost half a million inhabitants."[47] The civic secretary held frequent meetings with the health commissioner or his representatives, spoke with him repeatedly on the telephone, and kept up a cordial correspondence. Ruhland increasingly relied on the City Club to support health department expansions with investigations, recommendations, and appearances before budget committees and common council meetings. The City Club retained its independence, occasionally reaching conclusions critical of health authorities, but usually found that citizen interest was most often fostered by aiding health department work. The club, in the words of one of Milwaukee's mayors, was the "spark plug of many great local

---

[46] *Ibid.*, "copy" for the City Club *Bulletin*, Vol. 2, December 1916. See also the *Survey*, January 20, 1917. The agencies that contributed workers included: the Health Department, Associated Charities, School Nurses, Visiting Nurse Association, Children's Home Society, Hebrew Relief Association, Juvenile Protective Association, Central Council, Abraham Lincoln Settlement, Grand Avenue Congregational Church, First Baptist Church, Wisconsin Anti-Tuberculosis Association, Big Brothers, Big Sisters, and Washington High School. The members of the City Club Public Health Committee who organized the event were: Drs. J. Gurney Taylor, Hoyt E. Dearholt, Louis M. Warfield, Irene G. Tomkiewicz, S. G. Higgins, James A. Cavaney and A. W. Gray, Messrs. George Geetz and George Elwers and Miss Marion Ogden. See also the letters in the City Club papers planning the survey, Box 3 dated September through November, 1916.

[47] See the City Club Papers, Box 8.

movements," among which was the promotion of increased governmental activity for public health and welfare.[48]

The volunteers who engaged in public health reform in Milwaukee helped to bridge the distance between the health department and the masses of people in the early twentieth century and thus helped the city to overcome one of the major obstacles to public health reform that had plagued it in the nineteenth century. By beginning multiple small reforms, many of which were responsive to citizen input, the clubs and charities created an atmosphere of acceptance and understanding among Milwaukee's poorer and immigrant groups. They also helped the middle- and upper-class inhabitants to realize the importance of health reforms even when they cost a great deal of money. Thus by 1910 the city was prepared for a comprehensive government-run health program, its success assured by the great variety of people who were already committed to it: government bureaucrats, upper- and middle-class men and women who had staffed the voluntary organizations, civic-betterment professionals and business people whose future seemed dependent on a more effective health program, and the many working-class individuals who had benefited from the voluntary associations. Not dependent on a single political ideology or economic interest group, the city health program thrived. The integration of private and public health reform efforts helped Milwaukee to gain its reputation as one of America's healthiest cities in the twentieth century.

---

[48] Mayor Carl Zeidler was quoted in the inventory notes to the City Club papers in the WSHS. There is no evidence in the City Club Public Health Committee Papers of any communication between it and the Health Department during the 1914 "milk war" during which Ruhland certainly could have used some support. It might be that the school inspection issue, beginning in 1916, marked the first real working together of the City Club and the health department.

CHAPTER SEVEN

The Healthiest City

When Health Commissioner John P. Koehler accepted the large bronze plaque commemorating Milwaukee's top place in the Class I division of the first annual U.S. Chamber of Commerce and American Public Health Association Health Conservation Contest in 1930, he did so on behalf of his fellow workers in the city health department and also for "the local Association of Commerce, the private and public welfare, health, and educational agencies, and many others that have contributed their share." Koehler understood that the "great honor" bestowed upon Milwaukee resulted from years of concerted effort on the part of many diverse groups of people, inside and outside city government.[1] The city's outstanding health record rested on the shoulders of the health-reform efforts of the nineteenth and early twentieth centuries.

Although financial, technical, and even some political

[1] John P. Koehler, "Acceptance Address" following William Butterworth, "Inter-Chamber Health Conservation Contest," *American Journal of Public Health* 20 (1930): 635. See also Charles W. Gold, "Inter-Chamber Health Conservation Contest," *ibid.*, 22 (1932): 727-730. Milwaukee again won first place in 1932, 1936, and 1939, captured second-place honors in 1931, 1933, and 1935, received a Special Certificate of Merit in 1934, and was placed on the National Health Honor Roll in 1941, 1942, and 1943, after which date the national health contests were discontinued. See also Visiting Nurse Association Papers, Reel 1, p. 309, WSHS, and John P. Koehler, "Milwaukee Wins Again," MHD, *Bulletin*, June, 1932, pp. 2-4.

difficulties remained, twentieth-century public health programs found easier acceptance and generated less conflict
than those of the nineteenth century. The amalgamation
of public and private participation, the broad base of public
support, and the governmental commitment to promoting
and maintaining citizen welfare combined to smooth the
way for Milwaukee's health reform efforts. Two examples
from the decade after 1910 illustrate the new patterns. The
first, the Child Welfare Commission established in 1911,
exemplifies how the coordination of public and private efforts led to major health benefits; the second, the 1918
influenza epidemic, shows how Milwaukee approached the
familiar old problem of a raging epidemic. Both examples
reveal the ways in which the twentieth-century health department worked with the community to solve the city's
health problems, and they illustrate how and why the city

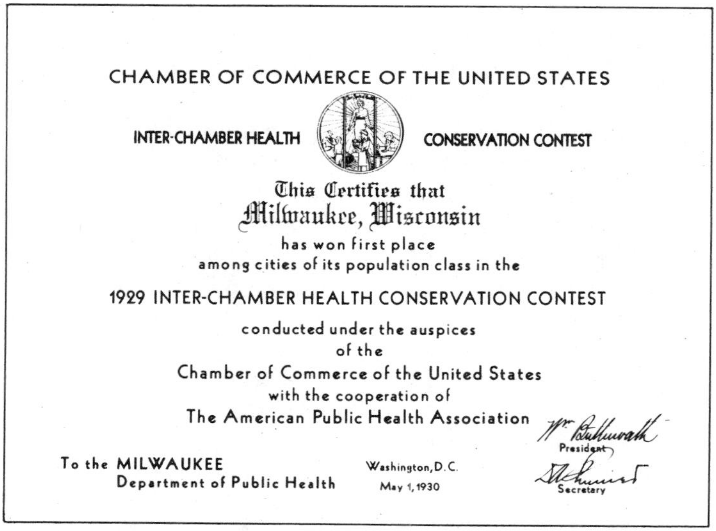

Figure 24. Placard commemorating Milwaukee's first place in the first
national health conservation contest, 1929. Courtesy of City of Milwaukee
Health Department.

of Milwaukee won the healthiest-city accolade repeatedly after 1930.

Health officials had long been concerned about high infant mortality. Prior to the twentieth century they had directed their attempts to increase milk controls specifically at saving Milwaukee's babies from unnecessary death. They had worked to decrease the dangers of contagious diseases through vaccination programs and epidemic control, focusing especially on children. Sanitarians and physicians had implemented school health programs—limited as they were—to save Milwaukee's youth from preventable or curable diseases. In addition to these public measures were the contributions of private agencies. Women's clubs, settlement houses, ethnic and religious benevolent associations, and physicians' groups had directed significant attention toward alleviating the suffering of Milwaukee's youngest citizens. They had provided food, milk, and clothing, sent the children to camp, and aided the sick and their families. Yet all the public and private efforts remained uncoordinated until the activities of Emil Seidel's Socialist administration, 1910 to 1912.

Ideologically committed to public responsibility for health and welfare, the Socialists, in their 1910 platform, promised to increase health facilities. Once in office they quickly voted to build a new city hospital. The rest of their health program was less clearly focused, and Seidel waited until the second year of his administration to develop a health program with an emphasis on children. The precipitating event was the arrival in April 1911 of a New York reformer, Wilbur Phillips, who came to Milwaukee to participate in the first Socialist government in a major American city. Phillips immediately pressed his enthusiasm for the potential health benefits of a coordinated child-welfare program upon the city. He believed that "ignorance and poverty" caused Milwaukee's high infant mortality and that both causes could be eliminated.[2] Phillips convinced Mayor Sei-

2 Wilbur C. Phillips, "Community Planning for Infant Welfare Work,"

del that the health dangers of childhood could be overcome through a systematic municipal program. "We know how to make timber into lumber and shape it into implements," said Seidel, agreeing that child health needed more attention. "[W]e know how to smelt ore into metal and manufacture the metal into the necessities of life, but we have not learned to properly handle our boys and girls."[3] Seidel supported Phillips' plan to mold good and healthy citizens through neighborhood-based, prevention-oriented, child-centered public health stations, and he put his authority behind the establishment of a demonstration district in which the method could be tested. Phillips also won the support of Republican Dr. John M. Beffel, Seidel's opponent in the 1910 mayoral campaign, and alderman Joseph P. Carney, leader of the opposition in the common council, both of whom helped to make this Socialist experiment a popular rallying point for people of all political persuasions. "It's going over," Wilbur confidently wrote to his wife-to-be Elsie Cole, still in New York, and he happily watched as the council appropriated $5,000 to begin the semi-municipal agency, whose budget was to be supplemented with private funds.[4]

Proceedings of the National Conference on Charities and Correction, 1912, pp. 40, 42 (hereafter cited *Proceedings NCCC*, date). Although I hesitate to ascribe all health advances in this period to the fact that the Socialists were in power, it is clear that Phillips would not have come to Milwaukee under any other circumstances. Certainly the child welfare stations would not have been adopted in 1911 if Phillips had not come, and the decentralizing emphasis on community prevention centers could not have developed when it did without this initial impetus. Thus, having a self-consciously Socialist administration molded the form of public health activity in this period, and, as the succeeding discussion shows, significantly influenced Milwaukee's twentieth-century health program.

[3] Seidel was quoted in the *Free Press*, April 30, 1911, from his speech to the mother's class at the Milwaukee Maternity Hospital. See also "The Milkman of Milwaukee," The *Survey*, 26 (April 1911): 171-172.

[4] Wilbur Phillips described the quick acceptance of his ideas in *Adventuring for Democracy* (New York: Social Unit Press, 1940), pp. 63-65. See also *Free Press*, May 14, 1911; Milwaukee Common Council *Proceedings*, file #272, May, 1911, pp. 77-78.

In addition to convincing Seidel of the benefits that the city would enjoy from focusing on children, Phillips met with the private agencies already involved in child welfare work to create a political groundswell for one centralized organization to coordinate all municipal efforts. "You have a wonderful opportunity here," he told the one hundred delegates from fourteen philanthropic agencies who gathered to support the proposed Child Welfare Commission: "I have found that there are many workers in Milwaukee, but many of them do not know what other workers are doing. A plan should be made to organize all the workers, and the work started in one section of the city. When it once becomes established, it will be an easy task to multiply, until the whole city is covered."[5]

As a first step toward city-wide coverage, Seidel appointed a Child Welfare Commission designed to win full community support. It included Beffel, whose name was already linked with anti-tuberculosis campaigns; William F. Fitzgerald, president of the Merchants and Manufacturers Association; Dr. Gustave Hipke, medical head of the Maternity Hospital; Dr. Lorenzo Boorse, a staff physician at Babies Free Hospital who had worked for clean milk supplies through the Milwaukee Medical Society; and Sarah Boyd, president of the Visiting Nurse Association and a successful businesswoman. The well-regarded group of citizens did not include a single Socialist, because Seidel did not want to endanger the future of the commission by appointing anyone controversial. A pragmatist, Seidel agreed

[5] *Free Press*, May 2, 1911. The agencies represented were the health department, medical department of the public schools, Central Council of Philanthropies, Wisconsin Anti-Tuberculosis Association, Infants' Home Hospital, Children's Free Hospital, Visiting Nurse Association, Infants' Fresh Air Pavilion, Milwaukee Medical Society, Medical Society of Milwaukee County, Children's Betterment League, Jewish Aid Society, Milwaukee Maternity Hospital, and the Social Economic Club. See also Carl D. Thompson, "Health Work in Milwaukee," *The Coming Nation*, August 19, 1911.

with Phillips that "no political party has a corner on saving babies." Yet, to insure that the social and political perspective out of which the idea grew did not get lost, the mayor appointed Socialists Wilbur Phillips and Elsie Cole Phillips, who had joined her new husband in Milwaukee, to share the staff positions of secretary and program planner.[6]

The Child Welfare Commission represented the wide spectrum of middle- and upper-class interests in the city's health-reform movement. Backed as it was by the Socialist administration and by a social theory that gave ideological significance to the small social experiment, the commission epitomized the new reform coalition that emerged in Milwaukee in the twentieth century. Although the Phillipses directed the particular events, they worked with the active support of a large segment of the advantaged people in Milwaukee.

The commission selected the beleaguered fourteenth ward for the site of the demonstration project, because the poor Polish immigrants who lived there crowded into small houses, produced proportionately the greatest number of the city's babies, and suffered the highest rates of infant mortality. The Phillipses' immediate task in their efforts to save the south-side babies was to gain the support of their new neighbors, who sixteen years earlier had shown their hostility to health authorities in the smallpox riots. The Phillipses, who spoke no Polish, sought the aid of community leaders. They recruited local doctors by offering to pay them for their services at weekly clinics in the child welfare station. They enlisted the local parish priest, Father Szukalski, to help them to win the confidence of the predominantly Catholic residents. In collaboration with the Visiting Nurse Association and with its financial assistance, the com-

---

[6] Phillips, *Adventuring*, p. 64; *Sentinel*, May 12, 1911; *Free Press*, May 12, 1911. The *Sentinel*, avidly anti-Socialist, was so in favor of the Child Welfare Commission that its editor exclaimed, "we are sorry it was not made more thoroughly representative by the inclusion of a socialist member" (May 13, 1911).

Figure 25. Original child welfare station on Milwaukee's south side, 1911. From MHD, *Annual Report*, 1912, p. 88.

mission hired three nurses to staff the station and to carry the message of good health to homes throughout the area. After months of planning, the child welfare station, designed to serve the thirty-three blocks of St. Cyril's Parish, opened on July 24, 1911. Father Szukalski encouraged his parishioners to take advantage of the free health services. To help the nurses to gain entrance into homes, he wrote a letter of introduction in Polish, which assured prospective clinic users that the health station "has the approval of the archbishop, and is entirely in keeping with my views. I . . . shall expect the mothers of my congregation to avail themselves of the assistance which the doctors and nurses will render."[7] He also endorsed the handbills that the nurses distributed throughout the community. Health reformers around the city shared the Phillipses' high hopes that the demonstration project would prove that Milwaukee's babies could be saved for approximately the same amount of money already spent to bury those who did not survive childhood.[8] The first day eleven mothers, seeking the doctor's advice, brought their babies to be weighed and examined. "The day has been a success," Phillips proclaimed, even while realizing the difficult task that lay ahead in proving the worth of preventive care to a community still suspicious of outsiders.[9] "Well begun," lauded the *Sentinel* editor, who glowed with chauvinistic pride as he noted that in this small section of Milwaukee "religious, racial, political and social extremes meet and clasp hands" in the new health station.[10] The Polish press shared the enthusiasm and welcomed the commission as a "very desirable" addition to the fourteenth ward.[11]

Through the summer and fall the station's nurses and doctors served south-side mothers and babies. Physicians led classes for mothers, teaching proper feeding and caring for infants and trying to stop the popular habits of giving coffee and beer to newborns to soothe their cries. One nurse remained at the health station each day, while the other two scoured the neighborhood, seeking new babies and "assuring a hearty welcome and free service" to all.[12] The work, while sometimes encouraging, was more fre-

---

[7] Father Szukalski's letter was printed in the *Sentinel*, July 24, 1911. See also *Free Press*, July 19, 1911.

[8] Phillips, "Community Planning," pp. 46-47.

[9] *Sentinel*, July 25, 1911. The *Journal* claimed that 8 mothers attended the first day, 12 the second, and 15 the third. (July 21, 1911.)

[10] *Sentinel*, July 25, 1911, editorial.

[11] *Kuryer Polski*, July 28, 1911. My thanks to William J. Orr, Jr., for finding and translating this editorial.

[12] *Journal*, July 29, 1911.

quently discouraging and slow because many residents remained hostile to suggestions from the health station. Phillips urged the nurses to make friends with the community women, but since two of the three spoke no Polish, communication remained difficult. As the Phillipses both realized, the community regarded all of them as "rank outsiders": "These women hadn't wanted us, hadn't known we were needed. Yet here we were, asking them not only to like our foreign manners and the general cut of our jibs, but to help us to do something we thought would be good for them. . . . We were in the uncomfortable, the silly and the futile position of tooting our own horn."[13]

Finally, with Father Szukalski's help, the Phillipses organized a mothers' committee of supportive neighborhood women to act as liaison between the health center and the community. These eight women sponsored evening socials, mixing "talk on babies" with refreshments and music to provide popular community entertainment.[14] Experience with reluctant fourteenth-ward parents convinced Wilbur Phillips that: "The mother must be won. This means a social program. . . . Recreation, amusement and fun must be furnished. Tea and cookies must be served. The pill of knowledge must be sugar coated."[15] By the end of the year the station boasted increased attendance at mothers' classes and baby clinics and a marked improvement in infant mortality in the district. Comparing infant deaths in the parish with the previous year's record, the commission claimed it had cut the rate from 12.5 per 100 live births to 4.4.[16]

Despite the dramatically improved health statistics, increasing community support, and the financial and moral

[13] Phillips, *Adventuring*, p. 72. See also Wilbur C. Phillips, "Psychology of Co-operation," *Proceedings NCCC*, 1915, pp. 62-64.

[14] Phillips, *Adventuring*, pp. 72-73. For more on community fear, see "Mothers Afraid to Let Babies Go To Pavilion," *Free Press*, July 23, 1911.

[15] Phillips, "Community Planning," p. 42.

[16] Common Council *Proceedings*, January 20, 1912, pp. 1181-1183; "Report of the Child Welfare Commission," MHD, *Annual Report*, 1911, pp. 178-183. See also Phillips, "Community Planning," pp. 44-45.

aid of city-wide reformers, the Child Welfare Commission found itself in trouble when the Socialists lost the 1912 municipal elections. The new "non-partisan" mayor, Gerhard A. Bading, physician and former health commissioner, sought to dissolve the semi-municipal agency in an effort to curtail government spending. Bading used his power as mayor to try to force the commission wholly into the health department, where its budget could be closely controlled and its activities monitored by physicians already on city salary. Ironically, the Child Welfare Commission had requested "surer legal footing" in its 1911 report and had always aimed for full city coordination, but its members had hoped to maintain and to expand existing personnel and activities. Bading was determined to get rid of Wilbur and Elsie Phillips, because, he said, they were not "residents" of Milwaukee. Bading's real complaint was the Phillipses' Socialist politics and allegiance to the regime he had just defeated. The confrontation threatened not only the child welfare project but the general health reform movement, which depended on cooperation among a wide group of reform-minded Milwaukeeans.[17]

Bading declared all-out war on the commission and, as a new mayor, elected by a formidable coalition of anti-Socialist Democrats and Republicans, he wielded considerable authority. But the commission and the Phillipses also had impressive support in the city. The battle lines were drawn when Bading insisted that the commission, once moved into the health department, be headed by a physician. Commission supporters thought that a sociologist who understood the social ramifications of instituting health changes—in fact, Wilbur Phillips himself—would be a more effective head. Phillips regarded the choice of primary agency as "comparatively unimportant," but he stood firmly against putting the Child Welfare Commission solely under physician control.[18]

[17] *Sentinel*, May 12, 1912; *Leader*, June 19, 1912.
[18] Phillips, "Community Planning," pp. 45-46, 47.

Through May 1912 and into June the common council debated the merits of both positions. Bading, who had efficiently cut health department expenditures during his tenure as health commissioner, hammered at the fiscal issues, but Phillips' supporters argued, as Dr. Lorenzo Boorse put it, that unless the commission continued in its present form, the city would pay "not in the few dollars that the administration is desperately trying to save, but in babies' lives."[19] Jennie Bernoski, head nurse at the welfare station, feared that the political controversy over the station would make the mothers, newly convinced of the merits of the project, "suspicious" again. She worried that: "Even the influence of Father Szukalski is not sufficient to counteract some of the rumors which are spreading through the district."[20] But when Bading insisted that a physician was "better fitted" for the work because he could save both babies' lives and city money, he convinced the council members, who voted on June 18, 1912, to put child welfare within the health department and under the supervision of a physician.[21] Elsie and Wilbur Phillips left town convinced that the decision would have the effect of "narrowing child welfare work" to doctoring the sick rather than allowing it to continue emphasizing preventive education.[22] Child wel-

---

[19] Quoted in the *Leader*, June 13, 1912. For a flavor of the debate, see, for example, *Sentinel*, May 13, 14, 28, June 7, 8, 18, 1912; *Leader*, June 11, 14, 17, 19, 1912.

[20] *Leader*, June 13, 1912. Her name was sometimes spelled Bernowski in the newspapers.

[21] *Sentinel*, June 8, 1912; *Leader*, June 19, 1912. See also Milwaukee Common Council *Proceedings*, file #2262, April 8, 1912, pp. 1,442-1,443.

[22] Phillips, *Adventuring*, pp. 106-114. See also Patricia Mooney Melvin, "Make Milwaukee Safe for Babies: The Child Welfare Commission and the Development of Urban Health Centers 1911-1912," *Journal of the West* 17 (1978): 83-93, who concludes, with Phillips, that the decision ended effective preventive child welfare in Milwaukee. It is true that their nascent idea of the social unit organization for a grass-roots community structure left Milwaukee with the Phillipses. But their ideas and goals for child welfare remained in the city. See also Patricia Mooney Melvin, "Neighborhood in the 'Organic' City: The Social Unit Plan and the First Com-

fare, reduced to routine health department business, seemed dead in Milwaukee.

But the Phillipses' analysis, the result of heated emotions following political defeat, proved fallacious. After getting rid of the Phillipses, the mayor allowed Dr. Emil T. Lobedan, the young physician put in charge of child welfare within the health department, to carry on their work and even to expand it. Bading realized that the large number of child-welfare supporters in the city, from the physicians and the women's clubs to the philanthropies and churches, wanted the work begun by the Phillipses to continue. Lobedan visited the demonstration project in the fourteenth ward immediately after Health Commissioner Kraft appointed him to the job, "found everything in good shape," and announced plans to open similar stations in other parts of the city.[23] Because of the broad coalition that supported it, the project formed under the Socialists withstood the political defeat of 1912 and paved the way for advances in child-welfare work in Milwaukee.

Lobedan followed through on his promise and opened a second child-welfare station before the year was out. The original station continued its work, staffed by the very same nurses and community physicians who had served under Phillips.[24] In the months from July, when Lobedan took over the work, to December 1912, the six nurses at the city welfare stations made 12,286 visits to patients. By 1913, with three stations functioning, the ten nurses then employed were making over 40,000 visits annually. Four years later Lobedan presided over six child-welfare clinics, whose nurses reported over 54,000 annual visits.[25] Throughout

munity Organization Movement, 1900-1920," unpublished Ph.D. dissertation, University of Cincinnati, 1978.

[23] *Leader*, July 11, 1912.

[24] Of the original seven physicians, identified as Drs. F. S. Wasiolewski, Irene Tomkiewicz, K. Wagner, Alfred Schulz, D. J. Dronzniekewicz, John Rock, and H. Gramling (*Free Press*, July 21, 1911), five continued to serve the south-side baby station. MHD, *Annual Report*, 1912, p. 89.

[25] E. T. Lobedan, "Report of the Division of Child Welfare," MHD,

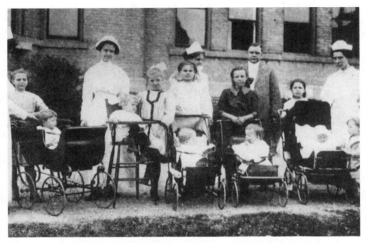

Figure 26. Nurses, children, "little mothers," and health officer in front of city child welfare station, 1912. From MHD, *Annual Report*, 1912, p. 95.

these years, Lobedan maintained Phillips' emphasis on prevention and education. The weekly physician-run clinics continued to weigh and examine babies and to give advice about infant care and feeding. In 1913 the nurses began "little mothers" classes for girls twelve to sixteen years of age to "spread the knowledge of the care and wants of infants" to future generations.[26] In the same year the child welfare division of the health department sponsored 400 public lectures on child care and disease prevention. The department's financial outlay increased from the original commitment of $5,000 in 1911 to $13,895 in 1913.[27] Lobedan continued to rely on the cooperation of "philanthropically inclined people" to help to provide fresh milk for the poor, sew appropriate baby garments, and staff fresh-air pavilions, hospitals, and state-fair exhibits.[28] In

Annual Report, 1912, pp. 87-95; 1913, pp. 112-127; 1914, pp. 31-42; 1915, pp. 79-92; 1916, pp. 30-37; 1917, pp. 75-87; 1918, pp. 22-24.

[26] MHD, *Annual Report*, 1913, p. 116.   [27] *Ibid.*, p. 118.

[28] MHD, *Annual Report*, 1914, p. 31; 1915, pp. 79-80.

1916 Lobedan expanded the department's focus to pre-natal care. Nurses visited expectant mothers and gave in-structions about "diet, exercise, fresh air, personal clean-liness, care of breasts, avoidance of alcohol and observation of urine and stools."[29] In 1920 the health department ran fourteen child welfare clinics and organized nutrition clin-ics in thirty-one schools around the city, giving "health talks and weekly weighings" to thousands of children.[30]

The work that Wilbur and Elsie Phillips began in 1911 proceeded apace after 1912 in the hands of non-Socialist health department officials who were able to continue and to expand it independently of political shifts in municipal government. Overcoming periodic interruptions in a way that had never been possible in the nineteenth century, health programs in the twentieth century achieved a sta-bility that permitted greater successes. The foresight of Wilbur Phillips and Emil Seidel in making the original Child Welfare Commission multi-partisan, in putting the care of babies above political squabbles, paid in health benefits for the children and for the whole city. Milwaukee's successful prevention program provided a useful model of centrally controlled but locally responsive health centers, a model that cities across America adopted.[31]

Milwaukee's experience with influenza attacks in the fall and winter of 1918 provides another illustration of how the city's previously established health reform coalition paved the way for rapid citizen mobilization and a favorable out-come. It also indicates that nineteenth-century techniques of coping with epidemics, when applied in the stable and

[29] MHD, Annual Report, 1916, p. 30.

[30] MHD, Annual Report, 1920, p. 14; Directory of Social Welfare Organi-zations: A Handbook for Citizens (Milwaukee: Central Council of Social Agencies, 1920), pp. 27-28; Gerald Burgardt, "History of the Child Wel-fare Clinic in Milwaukee," typescript of presentation in connection with the thirtieth anniversary of the Child Welfare Department, February 3, 1941. Copy in the Milwaukee Municipal Reference Library.

[31] George Rosen, "The First Neighborhood Health Center Movement—Its Rise and Fall," American Journal of Public Health 61 (1971): 1620-1635.

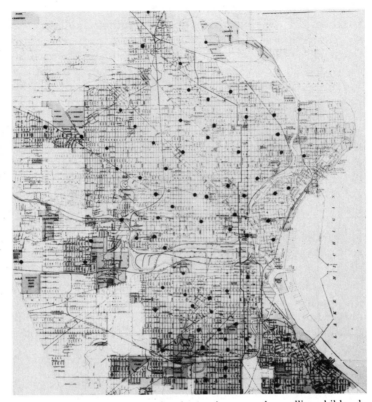

Figure 27. Map of Milwaukee showing stationary and travelling child welfare stations, 1932. Courtesy of City of Milwaukee Health Department.

relatively nonpolitical atmosphere of the early twentieth
century, could still be effective, especially within the context
of the newer emphasis on public education and inter-agency
cooperation. Because influenza struck suddenly and could
ravage large populations in a matter of weeks, responses
had to be prompt to be effective. Milwaukee minimized the
deadly effects of the disease by reacting quickly and boasted
one of the lowest death rates in the country.[32]

[32] *Sentinel*, January 8, 1919. See also Common Council *Proceedings*, file

228

Two "Jackies" from the Great Lakes Naval Training Station brought the Spanish influenza to Milwaukee in late September 1918. The health department, forewarned by the devastation in Boston, immediately began procedures to combat the disease. Health Commissioner George C. Ruhland requested that naval recruits be restricted to their base to minimize the spread of the flu from military personnel to civilians. He canvassed all Milwaukee physicians by telegram to ascertain the number of cases in the city and to request that they report all outbreaks to the health department, even though influenza was not on the list of reportable infectious diseases.[33] He consulted with leading physicians in the city to determine how best to "subdue the spread of Spanish influenza."[34] Most significant, Ruhland appointed four persons—two physicians and two businessmen—to serve as his advisory committee for the duration of the epidemic. The health commissioner thus involved in the decision-making process representatives of two of the influential groups affected by the impending epidemic. Ruhland would not repeat the mistakes that Kempster had made earlier by isolating himself from the community; he scrupulously took no action without consulting these four advisers.[35]

#15331, January 13, 1919, for the flu and pneumonia death rates for eighteen cities. There is no way of measuring the impact of health department activity on the course of the flu epidemic. The virulence of the strain and the naturally rapid course of the disease clearly were important to events. However, Milwaukee organized more efficiently than other American cities, and, whether or not this had a direct influence on mortality, the health department's command of the situation had the potential for increasing success.

[33] "How Milwaukee Organized Its Fight Against the 'Flu,' " *Wisconsin Medical Journal* 17 (1918-1919): 250-251: *Sentinel*, September 17, 19, 23, 25, 26, 1918; *Leader*, September 18, 21, 25, 1918.

[34] *Sentinel*, September 30, 1918.

[35] *Sentinel*, October 9, 1918; *Journal*, October 9, 1918; *Leader*, October 9, 1918. The committee consisted of Otto H. Falk, president of Allis-Chalmers Manufacturing Company, Carl Herzfeld, an owner of the Boston Store, one of Milwaukee's largest department stores, Dr. Louis F. Jermain, a prominent internist active in medical and public health activ-

The campaign that the health commissioner launched with the help of the medical profession, the business community, and numerous voluntary and religious organizations and with the full support of Socialist Mayor Daniel Hoan[36] utilized the traditional technique of fighting infectious diseases with isolation. Even though physicians could not cure influenza, they understood its microorganic transmission well enough to argue convincingly that isolation would limit the spread of the disease. Ruhland asked the common council for $15,000 to open isolation hospitals to care for the sick, and he immediately received the entire amount by an unanimous vote.[37] Simultaneously with this provision to care for the ailing, the health department mobilized an intensive advertising campaign to advise the public how to avoid contagion. Posters presenting precautionary measures appeared throughout the city; newspapers carried lengthy accounts of the situation and editorialized support for the health department efforts; churches and factories arranged for "four-minute" talks to advise listeners about the flu situation; and clubs postponed their meetings in the interests of flu prevention.[38] Ruhland held individual conferences with physicians, clergy, business people, theater managers, newspaper editors, and club representatives, who emerged from the meetings voicing their support for the emergency measures. The entire city prepared to resist the microbial offensive that threatened to become a greater menace to population than the Great War.

---

ities, and Dr. Hoyt E. Dearholt, executive secretary of the Wisconsin Anti-Tuberculosis Association.

[36] Although the Socialists did not hold a common council majority after their defeat in 1912, Socialist mayors served Milwaukee from 1916 to 1940 (Daniel Hoan) and from 1948 to 1960 (Frank P. Zeidler). Emil Seidel, defeated in his 1912 reelection bid for mayor, continued as alderman from 1916 to 1920 and from 1932 to 1936.

[37] *Sentinel*, October 10, 1918. Ruhland also received a matching $15,000 from the county.

[38] *Sentinel*, October 9, 1918; *Leader*, October 8, 9, 1918; *Journal*, October 14, 1918; *Free Press*, October 9, 1918.

During early October, case and fatality statistics revealed that precautionary measures were not curbing the spread of influenza. When the health department received reports of 256 new cases on October 7 and 340 more the following day, Ruhland realized that radical measures were needed if Milwaukee was to avoid severe devastation. In conjunction with the State Board of Health, and after conferring with his advisers, the Milwaukee health commissioner outlawed public gatherings until the danger of the flu passed: "I hereby order . . . that all theaters, movies, public dances, churches and indoor amusements and entertainments be hereby discontinued until further notice."[39] The day following the initial closing order, Ruhland added schools to his list of institutions to close and clarified his ruling as it applied to saloons: "crowds of men will not be permitted to gather in them and loiter about the interior of the place. [The ban] will not prevent a person from entering the saloon, purchasing a drink and then leaving the place."[40]

The health commissioner encountered some resistance from theater managers, who thought his order discriminatory. The managers noted that the ban permitted people to crowd into elevators, department stores, and streetcars, but not to enjoy the diversion of good entertainment. Clergy called attention to the illogical ruling that allowed people to enter a saloon to seek solace from the hard times, but not to pray together to the One who could alleviate suffering.[41] Ruhland carefully explained the purpose of his rulings to all who protested and warned that a wider ban would follow unless the city cooperated fully with the spirit as well as the letter of his directives.

More impressive than the occasional complaints was the widespread public cooperation elicited by this newest restriction on lives already limited by the exigencies of the war. From diverse groups of people Ruhland received praise for his efforts and full cooperation with his endeavors to

[39] *Sentinel*, October 11, 1918.   [40] *Sentinel*, October 12, 1918.
[41] *Sentinel*, October 24, 1918; *Journal*, October 21, 1918.

231

check the spread of the flu. The local Red Cross, pledged to emergency relief but already busy with war work, volunteered to help the city in its fight against flu. Its motor corps of 18 women collectively drove over 3,000 miles, answering 875 calls, during the worst three weeks of the epidemic. The organization's registration bureau coordinated the services of 156 nurses and nurses aides, who served in the public hospitals and in private homes.[42] The Visiting Nurse Association and numerous women's clubs helped to staff the three city isolation hospitals, which received hundreds of sick adults and children who could not be cared for at home. Individual women responded to the call for help and volunteered to serve the sick around the clock. Some opened their homes to the children of stricken parents.[43]

Hundreds of teachers, idle from their normal duties, canvassed the city, ferreting out unreported flu cases. Many helped in the hospitals and even did domestic chores, exchanging books for mops. One teacher wrote of her tasks at a city isolation hospital: "I went up to help in the kitchen. . . . I was directed to the sink full of dishes. Full? Piles, heaps of dishes! . . . There were several other teachers present, a business woman and a society girl, all working as busily as bees. Well, we washed and washed and washed, and boiled and boiled and boiled, and wiped—one odd shaped dish I know I wiped ten times during the night."

The teacher also fixed bottles for the babies, cooked dinner for the nurses, and helped with patient care. "She looked rather tired," observed a reporter, "but her eyes shone with enthusiasm and she laughed. 'It's a great life,' she said, 'if you don't weaken.' "[44] One hospital administrator commended "[t]he women of Milwaukee [who] have responded

[42] *Free Press*, November 12, 1918. For more on the Red Cross, see Richard Carter, *The Gentle Legions* (New York: Doubleday and Company, 1961), pp. 36-62.

[43] See, for example, *Sentinel*, November 10, 1918.

[44] "Fighting Flu is Real Work," by anonymous teacher, *Journal*, November 7, 1918.

wonderfully to the call and have worked in harmony at the most menial work."[45] For some, the sacrifice went beyond domestic labor; many of the hospital volunteers and staff succumbed to the flu as they tried to make other sufferers more comfortable.

Few Milwaukeeans found their lives untouched by the epidemic. Sickness and death abounded—the health commissioner estimated over 30,000 cases of flu in the last months of 1918—and ordinary life was disrupted daily by the closings and by the extraordinary efforts of the city to stem the tides of disaster. Businesses lost money by closing early and trying to dissipate crowds at their counters. All club life and gatherings stopped for over three weeks, and individuals, talking more on the telephones than ever before, tried to adjust to the demands made in the name of promoting the public good. "We are holding our own in the tug of war against influenza," boasted Dr. Hoyt Dearholt of the health commissioner's advisory committee, "because we have been pulling together and pulling hard."[46]

Indeed Milwaukee appeared to be conquering the flu. At the end of October, Ruhland, after consulting his advisers, decided to lift the ban on public gatherings. He credited "prompt action, an intensive publicity campaign unparalleled in the history of the city, and the responsiveness and co-operation of the public, both as official and unofficial organizations and as individuals" for the city's successful control of the epidemic.[47] On November 4 schools, churches, and theaters opened their doors after twenty-three days, people gathered together, and life in Milwaukee returned to normal. Ruhland warned that citizens should still be cautious, avoid crowds, and go to bed at the first sign of a cold. Milwaukee had lost almost 500 lives, but it had triumphed over the Spanish Lady.[48]

[45] *Sentinel*, October 17, 1918.   [46] *Journal*, October 24, 1918.
[47] "How Milwaukee Organized Its Fight Against the 'Flu,'" p. 250.
[48] *Sentinel*, November 2, 1918. See also *Proceedings* of the Common Council, October 21, 1918, pp. 553-554, and November 4, 1918, pp. 587-588, and

Unfortunately, the victory was short-lived. Near the end of November, Ruhland recalled his advisers and warned citizens that the influenza had returned. This time the public responded instantaneously. Billboards advising caution appeared overnight and full-page newspaper spreads warned citizens of the potential danger. One poster compared influenza to smallpox in order to alert the public. The health department placarded infected homes with bright red "INFLUENZA" signs and advised gauze masks for anyone who had to be out on the streets. Hospital representatives, Red Cross officials, business people, and aldermen met with the health commissioner to coordinate city efforts. Ruhland banned all public meetings and dancing, then closed schools and libraries, and imposed a half-capacity rule on theaters, churches, and stores. The order restricted seating in churches and theaters to alternate rows and prohibited other public gatherings altogether. The health commissioner threatened a complete ban—closing even retail stores—if people violated the partial ban. "It's up to the public," Ruhland

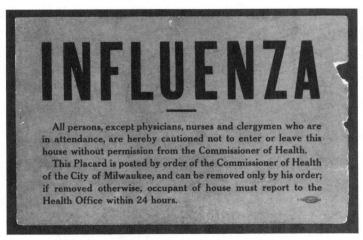

Figure 28. INFLUENZA Placard. Courtesy of City of Milwaukee Health Department.

announced. "Unless every man, woman and child does his or her bit to prevent the spread of influenza and obeys the orders of the city and state health officers, Milwaukee will be closed down tight next Tuesday."[49]

Department stores found it especially difficult to cooperate during the pre-holiday shopping season, but the health department posted inspectors in the major shops to keep crowds away.[50] Streetcars presented the major problem. The Electric Company complained that it was impossible to run a full complement of vehicles because many drivers were sick. Ruhland insisted that extra cars were essential to the public health, and he urged passengers not to enter the streetcars if they were crowded. Newspaper editorials and public opinion agreed that the Electric Company had "a positive public obligation" to furnish additional cars and drivers.[51] Even with sixty-three health inspectors posted at busy intersections to regulate the numbers of people boarding the streetcars, the company continued to violate the provision. One afternoon rider counted "fifty people standing in the aisles."[52] Another frustrated commuter wrote to the *Journal* editor: "The company seems in favor of guarding the health of the public, at least one would think so judging from the placards posted in the cars, suggesting that if a car is crowded, the patron take the next one. If one were to wait until the 'next car' that wasn't crowded came, he would never get downtown, for they are fifteen minutes apart and each is filled when it comes by."[53]

Other indications suggested that Milwaukeeans were finding the pre-holiday partial ban difficult to observe. Health

---

George C. Ruhland, "Influenza in Milwaukee," *Bulletin of the Milwaukee Health Department*, October-November 1918, pp. 3-4.

[49] *Sentinel*, December 7, 1918.

[50] *Sentinel*, December 11, 1918; *Leader*, December 16, 1918.

[51] *Journal*, December 9, 1918.

[52] *Journal*, December 13, 1918.

[53] *Journal*, December 15, 1918. See also the *Leader*, December 13, 14, 1918.

department officials investigated numerous reports of violations of the no-crowding rule at department stores and bowling alleys. The Boston Store, whose owner served on Ruhland's advisory committee, held a special sale on the "Wonderful Gyroscope," violating the rules by encouraging a crowd to watch the demonstration.[54] Even children, barred from department stores, complained. One wrote to the health commissioner: "Would you please give at least one afternoon or morning for just the children so they could buy Christmas presents. If you will do this, Dr. Ruhland," appealed Hope Albert, "you will make hundreds of children happy."[55]

But Ruhland never relented in his efforts to limit the spread of the flu: not for the businessmen, the streetcar magnates, or the children. Mayor Hoan stood behind Ruhland, who met daily with violators and supporters alike to impress upon everyone the seriousness of the situation. Ruhland's crusade ultimately had the support of the great majority of Milwaukeeans. Because they agreed that crowding exacerbated the spread of influenza and because Ruhland had spent much of his time in private conversations to convince diverse groups of people of the value of his campaign, Milwaukeeans tried very hard to cooperate. One store owner, who greatly valued his holiday trade, reluctantly agreed: "If lives will be saved by the closing of our shop we will close gladly."[56] Milwaukeeans generally cooperated by avoiding crowds, cancelling social functions, and wearing gauze masks in public. The benevolent agencies opened soup kitchens and provided volunteer help to the sick.[57]

Ruhland never needed to issue the full-closing order.

[54] *Leader*, December 16, 1918.
[55] *Leader*, December 21, 1918.
[56] *Journal*, December 10, 1918.
[57] See, for example, *Sentinel*, December 2, 3, 4, 5, 6, 8, 9, 11, 1918; *Leader*, December 3, 5, 6, 9, 10, 12, 1918; *Journal*, December 10, 11, 12, 1918.

Reports of influenza cases had declined sufficiently by December 25 to permit the health commissioner to lift the prohibition against large gatherings, although he insisted that holiday dancers wear six-layer gauze masks. This regulation remained in force through the New Year's celebrations, condemning parties to "resemble a band of holdup men 'from the neck up.' "[58] On January 2, 1919, schools reopened, and Milwaukee started the new year thankful that both the war and the epidemic had ended.

Milwaukee's experience with the flu in 1918 proved to be among the best in the country. Ruhland proudly presented the evidence to the common council: the city's death rate from influenza between September 14 and December 21 stood at a low 0.6 per 1,000 population; Minneapolis, which ranked second among large cities, had a record of 1.9. The health commissioner thought the low rate constituted a "justification of the methods used by his department to stamp out the disease."[59] He generously spread the credit: "I believe that the co-operation of the public and the willingness of doctors and nurses to work with the department, are largely responsible for the low mortality rate of Milwaukee. It was team work with the public, newspapers, Red Cross and other workers that helped us through two serious epidemics."[60]

Although such factors as climate, virulence, or population density may have contributed to Milwaukee's relative success, the city's health organization and community co-operation must be credited in part for the positive experience. In a very short time the health commissioner mobilized an army of volunteers, coordinated the efforts of numerous community organizations, plastered the city with educational literature, isolated the sick in their homes or

[58] *Sentinel*, December 24, 1918. See also December 23, 27, 1918; *Leader*, December 26, 1918.
[59] *Leader*, January 7, 1919.
[60] *Sentinel*, January 8, 1919. See also MHD, *Annual Report*, 1918, pp. 3-4, 8-9.

in city-aided hospitals, and assuaged the doubts of business people and politicians who feared personal loss from the emergency regulations. The central organization and community cooperation might not have been accomplished, even in wartime, unless previous efforts had paved the way.

The experience of other American cities confirms Milwaukee's superior organization in dealing with the epidemic. The health department of Newark, New Jersey, struggled actively against state intrusion during the influenza epidemic and was unable to bring order to the stricken city.[61] Philadelphia found itself floundering to cope with the thousands of sick and dying in need of care. Only the wartime Council on National Defense succeeded in bringing a semblance of order to that beleaguered city.[62] Alfred W. Crosby, Jr., who studied San Francisco's reaction to flu, concluded: "Institutional structure equal to the challenge of pandemic influenza didn't exist [in San Francisco] or was only just coming into existence as the pandemic passed its peak."[63] Milwaukee's public health structure, a governmental center laced with pathways to voluntary agencies and community members, born from the struggles of the nineteenth century and the amalgamations of the early twentieth century, provided a stable base on which the city could build its defense against influenza. Because it had developed a health structure open to community involvement and relatively unaffected by political differences, Milwaukee could successfully resolve the nineteenth-century dilemma of how to achieve public-health goals within the municipal political system.

Milwaukee was not, however, a perfect city in 1930 when it won the national health conservation contest. Many health

[61] Stuart Galishoff, "Newark and the Great Influenza Pandemic of 1918," *Bulletin of the History of Medicine* 43 (1969): 246-258.

[62] Alfred W. Crosby, Jr., *Epidemic and Peace, 1918* (Westport, Connecticut: Greenwood Press, 1976), pp. 70-90.

[63] *Ibid.*, p. 115. See pp. 91-120.

problems continued to baffle the health officers, and significant issues remained unresolved. Ruhland had not handled the milk war as efficiently as he handled the influenza epidemic, and he left a residue of bad feeling among milk dealers and suppliers that affected health department work. The city water supply, although treated chemically after 1912, continued to perturb drinkers; in 1918 it emitted an obnoxious odor that sent city engineers back to their drawing boards. Public health workers continue to this day to grope for solutions to environmental and chronic disease problems. Any suggestion that the twentieth century offered a simple progression toward enlightenment and progress would be misleading. Any suggestion that Milwaukee differed radically from its sister cities in health administration, honesty, or efficiency would be equally simplistic.

However, in the early years of the twentieth century Milwaukee did find a mechanism for dealing with public health concerns compatible with the city's history and ideology. Most significant, Milwaukee's health program worked. Officials maximized the city's health benefits by converting their idea of public responsibility into governmental administration with diverse community input and cooperation. This allowed the system to be responsive to the public's needs and to use community resources at the same time that the government effectively carried out its duties. The balance of political interests and community aspirations with centralized health department control allowed one of the nation's healthiest cities to emerge on the shores of Lake Michigan.

CHAPTER EIGHT

The Process of Change

The Milwaukee health department metamorphosed from a one-person, limited-budget operation to a professional, fully staffed organization in the sixty years following its creation in 1867. In the process of meeting the disease challenges of the rapidly expanding nineteenth-century city, health officers developed a clear vision of their purpose, learned how to manipulate the political system to achieve their goals, and gathered a fairly stable coalition of community support. The department adjusted to twentieth-century health problems, changing its operational focus from power confrontations to public education and its health interests away from environmental sanitation and infectious diseases toward preventive medicine and the management of chronic diseases. Through the years 1867 to 1930 the health department lost many of its politically controversial aspects and settled down to a routine existence within municipal government. It became part of the establishment that it once had challenged.

This book has traced the evolution of municipal responsibility for public health and welfare in Milwaukee. The components of urban reform revealed by the Milwaukee experience now must be analyzed. What factors promoted or retarded urban sanitary reform? Who was involved in health reform and for what reasons? How and why did local government come to accept greater responsibility? What role did individuals play in the reform process? Exploring answers to these questions will help us to understand the dynamics of public health reform in urban America. The ingredients of change and the relative importance of each varied over time and depended upon the specific health problem. The basic elements of health reform, however, can be identified in six categories: the nature of the issue, medical theory and abilities, economic interests, political pressures, individual actions, and social and cultural diversities. Municipal responsibility developed most readily when the health problem was acute, when medicine understood the crisis and could solve it, when business interests could profit from the changes, when at least one major political party accepted the reforms, when a strong but sensitive individual emerged to guide the reforms through the system, and when different cultural groups agreed on the medical solution. Obviously, no situation ever existed in which all of these criteria were met simultaneously; reform instead struggled in a setting characterized as much by resistance as by acceptance. The Milwaukee experience indicates that political and economic factors generally were most important in determining whether or not health reforms would succeed, but that on occasion each of the other elements could be paramount. While it is thus impossible systematically to rank the levels of importance of the various components of urban health reform, each must be assessed in turn to understand the process of change.[1]

The frightening and dramatic quality of the unexpected

---

[1] Other historians have analyzed some of the following components. See, for example, Barbara Rosenkrantz, "Cart before Horse: Theory, Practice, and Professional Image in American Public Health, 1810-1920," *Journal of the History of Medicine and Allied Sciences* 29 (1974): 55-73; James H. Cassedy, "The Flamboyant Colonel Waring: An Anticontagionist Holds the American Stage in the Age of Pasteur and Koch," *Bulletin of the History of Medicine* 36 (1962): 163-176; among others. Also, Joshua Ira Schwartz, *Public Health: Case Studies on the Origins of Government Responsibility for Health Services in the United States* (Ithaca, N.Y.: Department of City and Regional Planning and the Program in Urban and Regional Studies, 1977) explores many of these factors in his examination of the secondary literature.

provided the first impetus to health reform. Nineteenth-century urbanites learned to live in an environment fraught with health hazards, but when new crises erupted into the already bleak health scene, as they did repeatedly, many people found the added burdens unbearable. Periodic epidemics especially increased public awareness of the need for expanded health services. When smallpox or cholera threatened Milwaukee, citizens reacted vigorously. Not only were these diseases infrequent visitors, and therefore possibly preventable, they also carried ghastly symptoms and produced perilous outcomes. Officials trying to stem the tides of epidemic disasters received the public's cooperation at least while the crises lasted. Because of the fear generated at times of acute distress, epidemics frequently increased the power and authority of the health department. Conversely, chronic diseases, which killed more people than the epidemics, did not easily win the attention of citizen groups or health officials.[2]

The Milwaukee response to smallpox illustrates the typical pattern of increased interest and activity at the time of crisis. During five nineteenth-century smallpox epidemics, the health department cemented its authority to control infectious diseases. Only once, during the 1894-1895 episode, did this pattern not hold true, but even then increased attention produced greater than normal activity.

Similarly, the extremely foul odor of garbage brought sanitation to the attention of both medical and municipal authorities. Bad smells, according to the miasmatic theory of disease, indicated a medically dangerous condition. The smells of rotting organic material (as well as the stench from the factories trying to dispose of it) were so disgusting that

[2] Other historians have noted the different responses to endemic and epidemic diseases. See, for example, John Duffy, "Social Impact of Disease in the late 19th Century," *Bulletin of the New York Academy of Medicine* 47 (1971): 797-811; and Charles Rosenberg, *The Cholera Years: The United States in 1832, 1849, and 1866* (Chicago: University of Chicago Press, 1962).

they caught citizen interest and led to political action. Not surprisingly, the major pushes to eliminate the garbage menace came during the summer months, when rapid decay produced the worst stench.

The Milwaukee experience abounds with other examples that support the contention that unusual and acute disasters encouraged health reforms more than did the typical endemic problems. Tuberculosis, the city's major killer, received almost no attention until the turn of the twentieth century, in part because it was familiar, its symptoms lacked drama, and the disease took many years to kill its victims. The sewage clogging Milwaukee's rivers became an issue only when unusually hot summers exacerbated the concomitant smells beyond their usual state. Milk, which looked the same whether it came from inspected or uninspected cow barns, did not readily grip the public's attention. Citizens appreciated the need for water purification only after chunks of garbage poured out of their water faucets or when an epidemic of typhoid fever erupted.

Medical knowledge and ability delineated the boundaries of what was possible in health reform. When the state of knowledge changed, health reforms shifted direction. For most of the nineteenth century the filth theory of disease justified sanitary reform. Armed with this theory, physicians provided the necessary orientation for health policy, encouraging street cleaning and thorough disposal of organic wastes. During the last third of the nineteenth century, Milwaukee physicians gradually accepted the tenets of the germ theory, intellectually substituting bacteria for dirt in their understanding of disease causation and transmission. This cognitive switch within the medical community paved the way for changes in daily health department activity, although practical application lagged behind the intellectual acceptance of the new ideas. By the end of the century, public health physicians searched for new ways to combat disease that would be in keeping with their new ideas and began stressing specific bacteria and scientific analysis rather than general environment concerns. Medical inspections of individual schoolchildren replaced the earlier emphasis on sanitary school environments, although both remained important. Laboratory inspections of milk and water increased dramatically. In the twentieth century, physicians decreased their demands for clean streets and increased their search for sick individuals with specific communicable diseases. The new theories produced a different focus for public health activities but did not eliminate traditional practices.

During the transition from miasma to germ theory in Milwaukee, the daily activity of public health workers retained much of the old while incorporating some of the newer ideas. Because dirty environments contained elements that fostered bacterial growth and diffusion, physicians looking for bacteria frequently realized that basic sanitation campaigns could still aid their efforts. Wilbur Phillips' attempts to bring clean milk to infants and to improve cleanliness in the crowded homes of immigrants had direct precedent in earlier reform efforts. Physicians fighting epidemics of smallpox and influenza in the early twentieth century relied heavily on tactics that had been important to their miasma-believing forebears: isolation at home or in a hospital, vaccination (for smallpox), closing of public gathering places, and disinfection of personal belongings. Because specific disease preventives and cures that developed as a result of bacteriologic understanding were not immediately available, early-twentieth-century health departments continued to rely on familiar activities at the same time that they remained alert for new advances. Thus the health department readily adopted diphtheria antitoxin in 1894 and 1895, setting up culture stations, laboratory analysis of sputum, and searching for diphtheria victims, while continuing to work for general environmental sanitation.

Important to our understanding of the role of medical knowledge in public health reforms in the period from

1860 to 1930 is the multiplicity of possible solutions to most health problems that existed under both theoretical systems. Just as there were many ways available to the believers of miasmatic theory to sanitarily dispose of garbage, there were many suggestions for purifying milk available to believers of bacteriology. Both medical theories provided a range of potential health reforms. Under either theory, when medical opinion divided on what should be done to solve any particular problem—as it did, for example, about the efficacy of vaccination—the medical disputes helped to move public health problems into the political arena.

There was no single best way to dispose of growing mounds of urban wastes, and local officials, while accepting the medical necessity for cleanliness, found the doctors of little help in implementing the theory. City officials haphazardly tried all the various garbage disposal alternatives and became prey to various political and economic pressures before finally, at the beginning of the twentieth century, settling on cremation at a municipal plant. Medical uncertainty about the best solution complicated and slowed the urban response to the garbage problem and repeatedly allowed room for extra-scientific intrusions.

Medical knowledge did not provide a simple solution to the problem of contaminated milk even after most physicians had accepted the tenets of the new germ theory. Some physicians supported certifying milk to keep the nutritious elements of raw milk; others favored sterilization to destroy all living organisms. Some doctors preferred pasteurization, a process that destroyed only some of the bacterial growth; still others advocated tuberculin testing of cows. The Milwaukee milk war illustrates that political circumstances sometimes brought these diverse (but not necessarily mutually exclusive) medical opinions into conflict, pitting advocates against each other. Because of the lack of professional cohesion, medicine alone rarely solved a public health problem.

Just as there were times when the medical profession was

divided and confused about how to solve a particular problem, so there were many instances when the doctors took decisive action at important junctures to further the cause of health reform. Doctors in the Milwaukee Medical Society, armed with their new understanding of disease transmission in the early twentieth century, initiated school medical inspections when it became obvious that the health department could not accomplish the innovation alone. Similarly, physicians united behind the health commissioner in favor of vaccination, helping at a crucial time to silence the growing sentiment against the procedure. Organized medicine supported a city hospital, certified milk, fresh-air classrooms, and the health commissioner's control over the spread of infectious diseases. Individual private practitioners daily aided the cause of public health when they vaccinated their patients and reported infectious diseases to the health department.

The interests of private medicine and public health did not always coincide. When school medical inspections grew into school health-advice stations, the profession tried to reverse the trend toward public medicine. When the health department required the reporting of tuberculosis, many doctors opposed the policy, because they thought that it would destroy patient-doctor confidentiality and jeopardize their practices. Some doctors worked against the health department's vaccination policies and lent their support to citizens who resisted that procedure. Health reform could not have been accomplished without the cooperation and leadership of members of the medical profession; but the relationship between the profession and public health reform was complex.

Economic factors frequently influenced the fate of health reform to a greater extent than either the catastrophic nature of the health issue or its medical components. Public policies on environmental sanitation and epidemic control affected the businesses of Milwaukee, sometimes adding new burdens that business sought to avoid, other times providing protection against unwanted competition. The diverse financial interests represented in the expanding city added an influential voice to the determination of health policies.

Local pride and inter-city competition provided an economic dimension of support for health campaigns to clean the streets and to eradicate epidemic diseases. Milwaukeeans believed that Chicago and Minneapolis-St. Paul threatened their own economic progress and, trying to surpass their rivals, became intolerant of serious sanitation problems.[3] A prosperous Milwaukee could not rise from the piles of rotting garbage that supposedly caused high death rates. As one health department motto put it: "It is in health that cities grow. . . . It is in disease that they are wrecked."[4] The growing nineteenth-century city, still insecure about its future and trying to outdo its neighbors, proved fertile ground for public health activity.[5]

A potent economic pressure to increase governmental controls over health came from the business interests that stood to gain directly by the measures. There were profits to be made from health reforms, and enterprises scrambled for their share. When Milwaukee assumed responsibility for garbage collection and disposal, contractors avidly competed for the new jobs. The Wisconsin Rendering Company went so far as to commit felonies in order to keep the city contract.

Even the reforms that cost money found advocates among the city's entrepreneurs. Milk inspections and requirements for clean cowbarns added to production costs, but some

[3] For a discussion of urban rivalry between Milwaukee and its neighboring cities, see Still, *Milwaukee*, pp. 196-197, 328-329, 343-344.

[4] MHD, *Annual Report*, 1911, frontispiece.

[5] A corollary that partially negates this positive effect of urban imperialism is that cities competed with each other also to attract industry. They were reluctant to take action that would discourage new enterprises from coming to the city, such as strict environmental regulation. However, no industry would consider establishing itself in an unhealthy location.

milk dealers nonetheless supported the changes. They could pass on most of the increased cost to consumers, and they knew that the costly requirements would force the marginal dealers out of business. The small dealers, having no financial cushion, bitterly opposed the reforms. A natural alliance developed between health reformers and the larger business interests, both representing the middle and upper classes, whose continuing prosperity depended on the growth of the city.

Of all the economic factors that led to public assumption of responsibility for garbage in Milwaukee, the most significant was the inability of the private sector to handle large municipal problems. Because of the dominant economic ideology, the city initially turned to the private sector to solve its sanitary crisis. Yet, after repeated attempts to work within that system, Milwaukeeans felt betrayed by their own laissez-faire philosophy. The city encountered such corporate inefficiency and dishonesty that it ultimately accepted the extreme alternative of municipal ownership. Only after the press revealed the corrupt activity of the Wisconsin Rendering Company, and people learned the extent of its misdemeanors, did Milwaukee seriously consider the sweeping and expensive alternative of public ownership. Municipal responsibility in this case resulted from the failure of the private corporations, operating within the traditional realm of political patronage, to cope adequately with large-scale municipal sanitary and health problems.

Economic factors sometimes worked against major changes in city health policy. Reforms cost money, and for most of the nineteenth century the city budget was stretched beyond its capacity. Legislators, trying to avoid commitments to expensive projects, delayed as long as possible decisions about waterworks, sewers, and garbage plants. Aldermanic reluctance to implement new health policies anguished the health officers, who believed that the city should rank protection of citizen health as its highest priority. Many citizens themselves did not support expensive health innovations

248

because the health benefits seemed so distant from their lives. People who did not understand or accept the connection between filth and disease did not support public sanitation activity. Nor did citizens whose milk jumped in price without any discernible change feel inclined to support the health commissioner's preoccupation with whitewashing cowbarns or inspecting milk samples. If public health reforms put people out of work, as did some of the milk reforms, public support declined even further.

The depression of 1893-1897 had a significant influence on the public reaction to health reform. Its immediate impact came during the 1894 smallpox epidemic, when the public waged war against the health commissioner, Walter Kempster, who lost public support partly because of his lack of sensitivity to the desperation of the unemployed and poor. Despite this hostile effect of the juxtaposition of an economic depression and an epidemic, the depression hardships eventually aided the cause of health reform in Milwaukee. As historian David Thelen convincingly argues, the depression of the 1890s jolted people away from their old political loyalties, produced new perspectives on how to solve city problems, and promoted a new reform coalition among the people affected by the economic shifts.[6] Because of the hard times, a significant cross-section of people was ready to accept government actions to alleviate public health problems. Without the depression, such a constituency may not have developed, and without such a constituency, public health reforms might not have been successful.

Political factors equalled the economic ones as major determinants of the course of health reform. No single political party monopolized sanitary reform in the nineteenth century, but all politicians became interested in health when acute crises forced the issue into the public arena. West-

[6] David P. Thelen, *The New Citizenship: The Origins of Progressivism in Wisconsin 1885-1900* (Columbia: University of Missouri Press, 1972).

side aldermen of both major parties opposed the Merz rendering plant's inner-city location, because their constituents felt so strongly about the smells. Both Democratic and Republican administrations supported garbage reform, and citizens blamed both administrations for inefficiency in sanitary affairs. Both major parties grappled with the milk crises, the river nuisances, and epidemic emergencies.

Kempster's impeachment at the height of the 1894 smallpox epidemic reveals how strongly politics could influence the city public health efforts. The council impeached Kempster more for political and social reasons—his refusal to cooperate with party wishes and his haughty superiority—than for his medical theories. Even after Kempster recovered his job, the aldermen could not forgive his lack of political loyalty, and they never supported his programs with much enthusiasm.

The structure of the municipal political system sometimes hampered health policy reforms. Before Wisconsin passed its home rule bill in 1924, Milwaukee, like most American cities, had to depend on the state legislature to solve many of its internal problems. Without consulting the city, the state legislature in 1891 gave the Wisconsin Rendering Company authority to dispose of Milwaukee garbage outside the city limits, authority it did not bestow upon the city. In 1898 Milwaukee had to beg the state, over the dissenting votes of the Rendering Company aldermen, for permission to issue garbage bonds.[7] The enforced dependence on the state legislature meant that Milwaukeeans could not control or implement their own decisions. Even within the city, reforms met procedural obstacles. Complex bidding for city contracts stymied efforts to solve city problems and often left council members open to outside influ-

[7] For more on problems of eminent domain, see Clay McShane, "Municipal Ownership and the Wisconsin Supreme Court 1870-1910," unpublished seminar paper, University of Wisconsin-Madison, October 1970. On home rule, see Still, *Milwaukee*, pp. 377-378, 558-559.

ence. Repeatedly the aldermen advertised for garbage bids, received them, opened them, announced them, and then stopped to consider new bids, thereby invalidating the whole process. Aldermen elected for two-year terms did not have time to become adept at municipal procedures before the voters turned them out of office. New activity in the health department depended upon how well those in power learned to manipulate municipal red tape and to stay out of the reach of corporate interests.

The political event that most influenced the success or failure of public health reform was the biennial municipal election. Every two years, people seeking election or re-election to the common council or hoping for a seat in any number of city offices went to the hustings. The pre-election activity usually hastened the reform process by mobilizing and focusing energies and forcing politicians to take a stand. As the *Sentinel* satirized the 1898 pre-election flurry, whatever the unresolved problem, council members seeking re-election "all . . . joined hands and circled to the right."[8] Under the pressures of a local election, aldermen decided to move garbage disposal out of the city in 1892. The municipal garbage plant resulted from another such election-time pressure.

Citizen opinion as voiced in the political arena thus could be a potent influence on health policy. A number of factors influenced the public response, including the acute nature of the problem, the cost of the suggested changes, and the perceived impact on daily life. Sometimes sheer fatigue governed Milwaukee's political response to garbage. The city struggled with garbage for only twenty-seven years, yet people felt worn down by the intensity of the process. Health commissioners complained of the garbage "nightmare," council members resented the time they devoted to the subject, and residents generally cursed the ever-present

[8] *Sentinel*, March 18, 1898. The editor called "Just 'fore election be as good as you can be" an "aldermanic text."

251

menace. The city tried so many methods of garbage disposal and got so embroiled in corporate and political corruption that people became ready to accept any solution as long as it seemed honest and permanent.

In the nineteenth century, political ideology played a relatively minor role in influencing the course of health reform. No consistent political program for improving the city's health or coping with its recurrent problems emerged until the 1910 Socialist platform advocated free health care for all citizens and municipally owned institutions to maintain sanitary environments.[9] The Socialists' position on health did not win them the 1910 election, but consistent health policies did emerge in twentieth-century Milwaukee, in part as a result of the efforts started under their direction.

From the mid-nineteenth century Milwaukee had exhibited strong liberal sentiments, which emanated in large part from the post-1848 German refugees. Reform organizations after the 1880s, such as the Municipal League and other civic-betterment groups, followed this tradition. Developing alongside this middle-class reformist movement was a strong labor movement, divided before 1910, but Socialist in ideology. Because of the class and sometimes ethnic differences exhibited between these two traditions, they might have remained separate but for the political confusions and corruptions of the 1890s. The Bennett Law, outlawing German language in the schools, stirred traditional loyalties and brought previously opposed ethnic groups together. Under the severe economic stresses of the 1890s, municipal responsibility became a reasonable alternative for people who could not seem to control their own destinies. Because the dislocations affected both workers who lost their jobs and entrepreneurs whose businesses were threatened, the two groups began to understand that they

[9] Frederick Olson, "The Milwaukee Socialists 1897-1941," Ph.D. dissertation, Harvard University, 1952; Marvin Wachman, *History of the Social Democratic Party of Milwaukee 1897-1910* (Urbana: University of Illinois Press, 1945).

had many interests in common. The mayoralty of Democrat David S. Rose from 1898 to 1906 and again from 1908 to 1910 solidified this coalition sentiment. Rose's tenure in office brought patronage politics to its fullest extreme in Milwaukee, and newspapers almost daily revealed new scandals and the outrageous behavior of the incumbency. Those who wanted to institute good and honest government saw that only by gathering a broad coalition could they hope to reverse current trends.

The Socialists appealed to this growing segment of Milwaukeeans who realized that the old-style politics could not adequately address the city's problems and that new, efficient, and honest government was needed to bring the city into the twentieth century. Socialist candidate Emil Seidel, exuding honesty and promising a fresh response to city problems, offered a significant contrast to boss Rose. Seidel's victory in 1910, and the victory of Socialist aldermen across the city, resulted from the hope of the majority of Milwaukeeans that they could recast city government in a new mold. Many of the people who elected Seidel in 1910 did so partly because the party modified standard ideological positions and responded flexibly to local needs and partly out of desperation about the future of the city under traditional politicians.[10]

The Socialists, dedicated to expanding governmental responsibility, brought a pragmatic approach to health problems and willingly included all political groups in their developing health policy. The child welfare commission, with its support from middle-class conservative reformers as well

[10] This interpretation evolved from my reading of Thelen, *The New Citizenship*, Still, *Milwaukee*, Olson, "Milwaukee Socialists," and Sally Miller, "Milwaukee: Of Ethnicity and Labor," in Bruce M. Stave, ed., *Socialism and the Cities* (Port Washington, N.Y.: National University Publications, 1975), pp. 41-69. For business interest in this period see Robert Wiebe, *The Search for Order 1877-1920* (New York: Hill and Wang, 1967) and Samuel P. Hays, *The Response to Industrialism 1885-1914* (Chicago: University of Chicago Press, 1957).

as from labor-union Socialists, illustrated the success of coalition health reform politics. The health department's organizational abilities during the influenza crisis also demonstrated the city-wide acceptance of governmental authority to control public health. Commitment to social reforms, which had been slowly developing in Milwaukee through the end of the nineteenth century, burgeoned under the Socialists between 1910 and 1912 into a full-scale public policy, and, supported by a wide range of reform groups, continued to influence health department activity throughout the twentieth century, regardless of changes at the polls. Emphasizing public education, health commissioners after Kraft continued, with the help of long tenures in office, to develop health programs based not on power confrontations but on community cooperation. The accolade that came to Milwaukee in 1930 represented the culmination of efforts of the previous twenty years to put a high political priority on health.

Milwaukee residents outside of city government held an important key to health reform. Some, like James Holton, who called the city's attention to the garbage-contaminated water, acted individually; others worked in groups like the Municipal League, the City Club, the Woman's School Alliance, the settlement houses, and the press, to effect changes in health policy. Occasional citizen organizations, like the Anti-Stench Committee, formed to agitate about a single issue and disbanded when the problem was resolved. Whether their commitment was short- or long-term, people with political and economic clout noticed and worked toward solving health problems before the city took action. Swill children and their families or small individual milk dealers, being both unorganized and powerless in the city, rarely influenced the course of events. But prominent business people, without whom the city could not prosper, were an important political force. Not only did middle-class citizen groups support sanitary reforms through mass meetings or in baby stations, but in many cases they initiated them, as when the Woman's School Alliance instituted sanitary reforms in the city schools. The clergy often spoke out from the pulpit on health policies or actually helped to organize community efforts such as the demonstration child welfare project in the fourteenth ward. Without citizen arousal and commitment, cities would have moved very slowly, if at all, to effect urban health reforms.[11]

Women's groups were particularly effective in helping the cause of public health. A natural extension of women's traditional roles, urban housekeeping and protection of children became keystones in the activity of the new women's clubs at the end of the nineteenth century.[12] As women increasingly moved outside the home, they took up jobs that were most familiar to them. The Children's Free Hospital Association, the Woman's School Alliance, and the social settlements reinforced health department efforts to improve mortality statistics. Upper- and middle-class clubwomen provided much of the daily work force for urban cleanups and child welfare and brought the ideals of the reform movement home to their families. Women like Lizzie Black Kander devoted their energies to city betterment and, through social contacts, opened the doors to financial help and political support for the reform cause.

[11] For other examples of public health reform following this general outline of private to public, see George Rosen, *A History of Public Health* (New York: MD Publications, 1958), pp. 382-404 and Schwartz, *Public Health*, pp. 118-123. See also John Duffy, *A History of Public Health in New York City, 1866-1966* (New York: Russell Sage Foundation, 1974); Rosenberg, *The Cholera Years*; Stuart Galishoff, *Safeguarding the Public Health: Newark 1895-1918* (Westport, Conn: Greenwood Press, 1975).

[12] On women's clubs and their activities see Mary P. Ryan, *Womanhood in America: From Colonial Times to the Present* (New York: New Viewpoints, 1975), pp. 227-232; William O'Neill, *Everyone Was Brave: A History of Feminism in America* (Chicago: Quadrangle Books, 1969), pp. 107-168; Edith Hoshino Altbach, *Women in America* (Lexington, Mass: D.C. Health, 1974), pp. 114-121; and Sheila M. Rothman, *Woman's Proper Place: A History of Changing Ideals and Practices, 1870 to the Present* (New York: Basic Books, 1978), pp. 63-93. See also, for example, *The Housewives League Magazine* 1913-1920.

Figure 29. Milwaukee City Hall helps in public health
education campaign, 1931. Courtesy of City of Milwaukee Health Department.

Some citizen groups did not favor health reform. The
Anti-Vaccination Society directly opposed health policies.
The milk dealers' associations fought the new regulations
to clean cowbarns and milk. Certain immigrant groups rioted
to voice their disapproval of health policies. As effective as
some of these negative reactions were, organized citizen
efforts more generally encouraged increasing municipal
control over public health problems.

Full, dramatic newspaper coverage of the extreme nature
of health disasters often stimulated citizen support, because
it brought the dirty environments and high death rates to

everyone's attention. Newspapers of all ethnic and ideological persuasions, as both information carriers and opinion shapers, influenced public thinking on the prominent health issues.[13] The *Sentinel*, even when voicing only lukewarm editorial support, aided milk reform by widely publicizing the foul cowbarns. The *Leader's* support of the strikers during the milk war kept the controversy alive and helped to stall enforcement of the tuberculin test. All newspapers participated in the smallpox riots by keeping their readers informed of the details of the disturbances and editorially favoring or opposing Kempster. Such heavy media involvement allowed ordinary citizens to maintain high interest in evolving health policy.

Individual health commissioners obviously affected the course of health reform during their tenures in office, but how much is difficult to assess. Because the job of chief health officer remained under political influence in Milwaukee until the twentieth century, physicians in this public office tried to oblige the mayors who appointed them and to appease their particular political party. Thus, to some degree, health commissioners were only carrying out the political interests of their administrations. When health commissioners, like Kempster in 1894, did not follow party tenets, they found implementation of their work particularly difficult. Why did no "Milwaukee Moses" arise to singlehandedly conquer the political and economic obstacles to health reform and bring health and order to the city?[14] Most of the people who held the job were not only able

---

[13] The *Sentinel* followed the Municipal League policy in the 1890s and supported German-speaking Republicans for local office. One of its greatest editorial stands was for clean government. The *Daily News*, the working-class newspaper, and the *Journal*, Democratic, offered good counterbalances to the *Sentinel*, as did the *Leader*, the Socialist daily. The *Vorwärts*, the Socialist paper preceding the *Leader*, was also ideologically outspoken; the *Abendpost* provided a German-Republican viewpoint, and the *Kuryer Polski* the Polish-Democratic. A full analysis of the role of the press in urban Progressive reform is needed.

[14] *Sentinel*, January 17, 1892.

doctors but were among the most successful and prominent in the city. But their medical abilities did not necessarily parallel their political abilities. Individuals like Walter Kempster fared more poorly with the political demands of the job than did Johnson or Wingate, who, although Kempster's equals among the medical elite, were more adept at manipulating economic and political interest groups in the city.

A comparison of any of Milwaukee health commissioners with Hermann Biggs, one of New York City's foremost health leaders, or with Charles V. Chapin, Providence, Rhode Island's health champion, reveals that none of the Wisconsinites had the Easterners' political foresight and purpose. Long tenure in office obviously helped Biggs and Chapin to win many of their goals, but their ability to work within the political system and their agility at mollifying antagonistic groups surpassed the talents of the best that Milwaukee had to offer. But Biggs and Chapin seem more anomolous than Wingate, Martin, or Wight: one can point to only a handful of super-successful health officers in late-nineteenth-century American cities. Most, like the Milwaukee group, functioned adequately but seldom masterfully. In the hands of people whose talents lay more with medicine than with politics, health reform frequently ran aground in the public waters of city government.[15]

The importance of political ability and public style is evident in a comparison of two Milwaukee reformers, Walter Kempster and Wilbur Phillips. Kempster's vision of a professionally staffed, laboratory-equipped health department, which incorporated some of the old techniques of sanitary cleanups with the newer germ theory, collapsed

[15] The effect a politically adept health officer could have on public health affairs in a big city is illustrated in Daniel M. Fox, "Social Policy and City Politics: Tuberculosis Reporting in New York, 1889-1900," *Bulletin of the History of Medicine* 49 (1975): 169-195; and James H. Cassedy, *Charles V. Chapin and the Public Health Movement* (Cambridge: Harvard University Press, 1962).

during his administration in the morass of city politics, in large part because he did not provide the necessary leadership and sensitivity to see it through. He was unable even to execute an effective response to an epidemic emergency, a task that had been accomplished by all of his predecessors.

With the advantage of hindsight, we can identify Kempster's major stumbling blocks. The economic depression of 1893-1897 would have created difficulty for any health administrator, especially for one trying to implement new hiring policies based on merit instead of on political qualifications. The shortage of jobs emphasized the ethnic and class divisions already present in the community. More important than the political and economic environment of 1894, however, were the methods that Kempster chose to try to overcome the difficult situation.

Kempster could have had two effective allies in his fight against smallpox. First, he could have cultivated his fellow Republicans in the common council by utilizing their criteria in appointing at least some of the temporary health department helpers—the quarantine guards, extra ambulance drivers, and hospital aides. Even while keeping to his goal of merit qualifications for the regular appointments, Kempster could have compromised on the emergency workers and bought some loyalty in the manner traditional to government officials. His principled position on patronage cost him important political support within the common council and on the streets.

Kempster's second potential ally was the medical community, particularly the doctors on the south side, the most severely afflicted part of the city. The health commissioner needed the south-side doctors if he was to be effective in vaccinating and isolating the sick. Yet he did nothing actively to cultivate this constituency. None of his assistant health commissioners came from the south side, nor did Kempster make any effort to hire or consult doctors from that region. Although he had much support from the medical community in his fight against smallpox, a large group of south-side doctors joined Emil Wahl in confronting him and in bringing him down. There were some disagreements within the profession with regard to smallpox treatment and prevention, but the division of physicians on geographical lines indicates that Kempster himself precipitated the split when he ignored the south-side doctors within the health department structure.

Kempster's basic flaw as a health reformer was his insensitivity to cultural differences. He either did not recognize or refused to admit that people in Milwaukee held beliefs that did not correspond to his own. This was clearly revealed in his inflexibility with regard to the law. He insisted on enforcing his interpretation of the health laws, which gave him the power to remove people forcibly to the hospital, because he was sure that he was right and that his rightness was self-evident. To his mind, children from crowded immigrant homes simply could not be isolated adequately at home. Yet Milwaukeeans of Polish and German descent believed the forcible-removal law to be inhuman and unnecesssary to health as well as contradictory to their traditions. These people thought that the sick deserved to be loved and cared for within their familiar rooms, not to be cast out of their homes into a strange place where their language might not be understood and where they would receive only the limited attention of strangers.

Kempster did not pause to consider possible alternative interpretations of the law, which might allow children to remain at home under strict quarantine, because he did not recognize that as his job. "I am here to enforce the laws," he insisted. "The question of the inhumanity of the laws I have nothing to do with." He was willing to "break heads" to get his way, but he was not willing to try to understand the strong immigrant resistance to his policies, to try to fathom why people who spoke a different language, whose lives had been uprooted by the move to this country, and whose lives were still riddled with economic and cultural dislocations, would think differently from himself, with his polished command of the language, his privileged position in society, and his powerful position within city government. Kempster and the new immigrants were literally from two different worlds. It is not surprising that they saw the smallpox crisis from such different class, cultural, and political perspectives. Kempster was unable and unwilling to try to bridge the gap.

Wilbur Phillips, in contrast to Kempster, designed his health reform program specifically to be politically successful and to build bridges between the reformers and the immigrants. Easily winning the support of his own Socialist party, he worked very hard to convince other people that child welfare work was essential to solving the major problem of higher-than-necessary mortality. He rallied women's clubs, professional organizations, and individuals from all political parties to his cause. Within a month of coming to Milwaukee, Phillips had a broad-based constituency.

Phillips clearly understood his role as an outsider in a heavily Polish immigrant parish. He aimed to win acceptance for the health center in order to build a community base for instituting preventive health activities. Phillips' sensitivity to his "otherness" in St. Cyril's Parish allowed him to avoid some of the major pitfalls that Kempster had encountered. Phillips went directly to the south-side doctors—some of them ethnic members of the community—and sought their cooperation. He also sought the aid of the parish priest, the local midwives, and nurses trained to be sensitive to cultural differences. He consciously tried to minimize the image of the health center as alien and to encourage immigrant women to bring their children for preventive checkups. Whether or not he knew it, Phillips implemented a program that reversed most of the shortcomings of Kempster's approach sixteen years earlier. Kempster's and Phillips' ideas about the necessity of clean environments, good food, and medical oversight were similar, but the methods of achieving their goals rested on antithetical premises about human differences and the role of authority.

Because of Kempster's political ineptitude, his medical program suffered; because of Phillips' political capability, his medical program flourished. Health officials might have learned from Kempster's mistakes, but they learned more from Phillips' successes. Community health centers expanded after Phillips left town and continued to incorporate his emphasis on local needs and resources. The advantages of working within ethnic communities to develop health education and prevention programs instead of Procrustes-like trying to impose an outside model on them became evident to health department officials as a direct result of Phillips' work.

The ethnic divisions in Milwaukee's population that confronted Kempster and Phillips severely challenged all health reformers. Because the ethnic differences frequently coincided with class differences, a wide gulf separated the occupants of the health office, chiefly native-born and English-speaking middle- and upper-class men, from the bulk of the population. The issue of infringement of personal liberties illustrated the kinds of problems raised for health reform by these divisions. Many immigrants who came to America to escape political oppression found any governmental authority over their private lives offensive. These liberty-seeking people felt that their rights to privacy were violated if health officers placarded their homes. If officials further wanted to invade their bodies with vaccinations, the newcomers protested even more strongly. When the health department outlawed public funerals, and when English-speaking government officials forced their way into immigrant homes to take smallpox victims to the isolation hospital, immigrant outrage knew no bounds. A government that could forcibly separate children from parents was no less oppressive than those the immigrants had fled by coming to America. Official arguments about control of mass infections and protection of the public health had little meaning for people who felt so personally threatened. The more adamant officials became, the more strongly they drove a wedge between themselves and the very people they sought to protect. Epithets hurled at Kempster in 1894 revealed how strongly the south-siders felt these class and cultural differences. The diverse traditions created barriers that led the public doctors to despair that any real change could occur in the crowded parts of the city.

The cultural divisions also reveal an element of social control that may have motivated some of the people advocating health policy changes. Reformers worked to alter habits born in the old country, especially those concerning diet, sanitation, and living conditions, to make them match an American middle-class norm of moderation, cleanliness, and order. The purpose was health; the method, assimilation and alteration of cultural patterns. Kempster tried to force compliance to the new ways; health commissioners by John Koehler's time accepted the differences and tried to educate the people away from their traditional habits. Either way, immigrants perceived many public health innovations as threats to their traditions and as attempts to change and control their lives. In Milwaukee, the city with proportionately the largest immigrant population of any in the United States, these cultural differences significantly influenced the course of health reform. In all American cities the varied interpretations of official measures to curb disease complicated the implementation of new health policies.

The clash of cultures evident in Milwaukee's experiences illustrates one of the basic paradoxes of turn-of-the-century reform. Reformers could not bring change and health improvements to the lives of urbanites without altering the structure of those lives. Reformers could not have cleaned up the Milwaukee milk supply without harming the small producers and raising the costs to the consumers. Infant mortality could not have been reduced without imposing change upon the lives of the immigrants. The diverse cultures of American cities could not have learned to live and work together without homogenizing some of their differences. Improving the health standards in the cities meant denying cultural identity, economic independence, and political choices to some people. Health reform, even when successful (maybe most especially when successful) cannot be tallied on simple mortality graphs, but has to be understood within the complex social and cultural milieu in which it struggled. Its benefits in terms of lives saved has to be balanced with its debits in terms of lives changed, diversities forgotten, freedoms squelched, livelihoods denied. Progress does not only move forward.

Chronological Outline of Public Health History in Milwaukee

1837 Milwaukee incorporated as a village
1843 Smallpox epidemic
1845 Milwaukee Medical Association formed
1846 Milwaukee incorporated as a city consolidating three sections
    Smallpox epidemic
1849 Cholera epidemic
1850 Population: 20,061
    More than one-third of Milwaukee's population German
1860 Population: 45,246
1862 Anti-hog ordinance
1864 Citizen petition to state legislature to establish an independent Board of Health
1867 Permanent Board of Health established by state legislature
    Dr. James Johnson appointed President of the Board of Health
1868 Smallpox epidemic; vaccination campaign
    Passavant Hospital accepts city smallpox patients
    Schools closed due to smallpox in second, sixth, and ninth wards
    E. S. Chesbrough survey for city water works
1870 Population: 71,440
    Death Rate: 20.93
1871 Appropriation to begin construction of Chesbrough's water works
    Sewer pipes laid

Milwaukee City dispensary opened
Smallpox epidemic; vaccination campaign
1872 Smallpox epidemic hits German population most severely
61 percent of all deaths under five years of age
1874 Health officer attempts slaughter house regulation
City water works opened
1875 "Swill children" replaced as garbage collectors by ward contracts
1876 Smallpox epidemic, hits immigrant Poles and Germans hardest
Push for city hospital
Wisconsin State Board of Health established (tenth in nation)
1877 Dr. Isaac H. Stearns appointed health officer
Prohibition on cutting ice below sewers
Health officer requests asphalt pavement for health reasons; ignored. Milwaukee streets paved with wood
Housing recommendations ignored by council
Purchase of land in eleventh ward for city hospital
Placard ordinance
Vaccination required for children to be admitted to public schools
1878 Dr. Orlando W. Wight appointed Milwaukee's first health commissioner
Board of Health becomes department of health
Health survey of public schools
Garbage contract let for whole city; disposal by feed and land fill
General milk ordinance prohibiting sale of impure milk; no enforcement provision
1879 Public suspicion of contamination in water supply
River nuisance
Night scavengers licensed and inspected by health department
City Isolation Hospital opened; no sewer or water connections
Crisis in garbage collection; six weeks with none
Survey of dairies conducted by health department
Milk ordinance rejected
1880 Population: 115,587
Death Rate: 20.68

Two assistants added to health commissioner's staff
Restrictions put on slaughtering process
Report of expert engineers—E. S. Chesbrough, George
Waring and Moses Lane—submitted on intercepting sewers

1881 Dr. Robert Martin appointed health commissioner
Smallpox "scare"

1882 Milk survey conducted by health department; *Sentinel* exposes foul conditions of urban dairies

1883 Milk legislation fails to pass

1886 Health commissioner collected and disposed of city garbage; dumped in Lake Michigan

1887 River nuisance leads to appropriation for Milwaukee River
Flushing Tunnel
Contract let to cremate city garbage
Milk ordinance passed to inspect milk and license vendors

1888 Ordinance restricts cattle slaughter to proscribed limits
American Public Health Association holds annual meeting
in Milwaukee
Milwaukee River Flushing Tunnel completed
Storing of pure and polluted ice in same ice house prohibited
Compromise milk ordinance passed requiring registration
in lieu of licensing; no inspections
Milk price jumps from 5¢ to 7¢ a quart

1889 Merz rendering plant disposes of city garbage

1890 Population: 204,468
Death Rate: 18.33
Dr. U.O.B. Wingate appointed health commissioner
Garbage "temporarily" dumped into Lake Michigan

1891 New water intake
Health commissioner recommends school medical inspections; no action
Milwaukee Anti-Vaccination Society established
Milk ordinance provides for licensure of milk dealers, inspections, and outlaws sale of swill milk

1892 Butchering within city proscribed; required to connect to
city's sewer system
Privy vault construction limited and regulated
Johnston Emergency Hospital opened
Cholera threat

Health commissioner given power to remove to hospital anyone suffering from contagious disease who is dangerous to the public health

"Garbage campaign" led to contract with the Wisconsin Rendering Company to operate Merz plant in Mequon

1893 City Isolation Hospital remodelled
Wisconsin College of Physicians and Surgeons opened
State law provided for appointment of Registrar of Vital Statistics

1894 Dr. Walter Kempster appointed health commissioner
Smallpox epidemic; riots
Diphtheria antitoxin available in city
Repeal of forcible removal ordinance
Milwaukee Medical College opens

1895 Dr. Kempster impeached and removed from office
Dr. H. E. Bradley appointed acting health commissioner
Diphtheria stations around the city distribute free antitoxin

1896 Dr. Kempster reinstated as health commissioner
Five assistants and food analyst added to staff
Laboratory begins functioning systematically
Public school nuisances corrected
Daily inspection of milk samples begins
Gridley Dairy pasteurizes milk
Civil Service legislation governs health department appointments

1897 State Supreme Court decision, *Adams v. Burdge*, limited authority to vaccinate schoolchildren without state legislation
Garbage contract with Wisconsin Rendering Company expired; garbage crisis results
First vote on municipal ownership of garbage disposal plant passed

1898 Dr. F. M. Schulz appointed health commissioner
Bacteriologist and chemist added to staff
Second vote on municipal ownership of garbage disposal plant passed
Strong milk and cream regulation ordinance fails to pass

1900 Population: 285,315
Death Rate: 13.88
Privy vault construction prohibited on streets having water and sewer pipes

Health department began medical inspection of schools to control spread of infectious diseases (temporary)

Ordinance regulating sale of horse and dog meat and sale of exposed food

1901 City Isolation Hospital #2 opened for diseases other than smallpox

Governor vetoed bill providing for compulsory vaccination

Contract let for building municipal garbage plant

1902 Jones Island Crematory, city owned and operated garbage incineration plant, opened

1903 Milwaukee Medical Society Milk Commission formed

Straus Depot opened in Milwaukee under auspices of Babies Free Hospital Association

1904 Movement to build a new city hospital

Anti-spitting ordinance

Smallpox epidemic

Certified milk sold at 14¢ a quart

Milwaukee County Medical Society appointed Tuberculosis Commission (precursor to the Wisconsin Anti-Tuberculosis Association, 1908)

1905 City water contaminated; boiling urged

Milwaukee Medical Society urged systematic school medical inspections

1906 Dr. Gerhard A. Bading appointed health commissioner

Bad meat scandal; meat markets to be licensed

Tenement law restricted ill-ventilated living conditions

City Building Inspector condemned City Isolation Hospital #2 as a fire trap

Milwaukee Medical Society executed trial medical inspection of schools

State legislation required vaccination of schoolchildren during an epidemic in their district

Bacterial inspections of milk begun

1907 Kinnickinnic Flushing Tunnel opened

Rudolph Hering report on garbage disposal in Milwaukee

Ordinance passed requiring milk to be bottled and sealed before sale

Visiting Nurse Association incorporated

1908 Ordinance on tuberculin testing of cattle passed; not enforced until 1926

1909 School medical inspection officially began under the School
Board
Common Council adopted Hering's plans for garbage disposal
1910 Emil Seidel elected mayor with Socialist majority in the
common council on platform that included free medical
care for all
Population: 373,857
Death Rate: 13.90
78 percent of population foreign-born or foreign stock
Erie Street Garbage Incineration Plant opened; Jones Island Plant closed
Dr. F. A. Kraft appointed health commissioner
1911 Rendering within city limits prohibited; ordinance not enforced
Sewerage Commission report found city's water contaminated; filtration plant recommended
Construction begun on new city hospital
Blue Mound Sanatorium for tuberculosis patients
Medical inspection of private schoolchildren instituted under the health department
Responsibility for garbage collection and disposal shifted
from the health department to the department of public
works
Approximately 50 percent of Milwaukee's milk pasteurized
Milk plants scored by health department—only two receive
"excellent" rating
Attack on "Hokey Pokey" ice-cream vendor
Child Welfare Commission opened fourteenth-ward demonstration project for free consultation on the care and
feeding of infants
1912 City water treated with hypochlorite of lime
South View Hospital (new city isolation hospital) opened
Division of Tuberculosis created within health department
Child Welfare Division created within health department
Gerhard Bading, former health commissioner, elected mayor,
non-partisan
1913 *Adams v. Milwaukee*, United States Supreme Court sustained
tuberculin testing of cattle, also giving the city jurisdiction
outside its boundaries
City financially supported Babies Fresh Air Pavilion

1914 Smallpox epidemic
Dr. George C. Ruhland appointed health commissioner
"Milk war" led to compromises on tuberculin test enforcement
1916 Chlorinating apparatus breakdown led to referendum for sewage treatment plant
Housing survey marked beginning of campaign against unhealthy housing
City Club Sickness Survey showed 10 percent of Milwaukeeans sick
Old Isolation Hospital razed
Pasteurization ordinance on all milk but certified
Ordinance prohibiting people with communicable diseases from working in establishments where food was prepared
Tuberculosis diagnostic clinics opened by health department (formerly under auspices of Society for the Care of the Sick)
1918 Flu hits Milwaukee; successful campaign launched to fight it
1919 School medical inspections unified for public and private schools under the health department
Health department decentralized with sub-stations on the north side and the south side
"Care of Baby and Young Child" pamphlet distributed to the homes of new-borns
1920 Pasteurization upheld in *Pfeffer v. City of Milwaukee*
Population: 457,147
Death Rate: 11.6
Venereal disease division created within health department
Visiting Nurse Association started its maternity service
1921 Dental Clinic opened by health department
Anti-noise ordinance
1924 Dr. John P. Koehler appointed health commissioner
1925 Virulent smallpox epidemic arrested by vigorous health department action; 427,959 people vaccinated
1926 Intensive diphtheria prevention campaign began using toxin-antitoxin and the Schick test
Tuberculin test enforced for cows producing for the Milwaukee market
1930 Population: 578,249
Death Rate: 9.6

All milk, except certified milk, pasteurized
Certificates of Merit issued by health department for eating
establishments
Milwaukee awarded first prize for cities over 500,000 population in national health conservation contest
1931 Milwaukee received second place in national health contest
1932 Milwaukee won first place in national health contest
1933 Common council borrowed WPA money to construct a
water purification plant
Milwaukee won second place in national health contest
1934 Scarlet Fever immunization program
Milwaukee received Special Certificate of Merit in national
health conservation contest
1936 Milwaukee again awarded first prize in national health contest
1938 County Medical Society, Visiting Nurse Association, and St.
Joseph's Hospital began community program for the care
of premature babies
1939 New city water purification plant opened
Health department evening venereal disease clinic begun
Milwaukee won first prize in the national health contest for
the fourth time
1940 Population: 587,472
Death Rate: 9.5
Dr. Edward R. Krumbiegel appointed health commissioner
1941 Milwaukee placed on the National Health Honor Roll in
the national health contest
1942 Pamphlet "Baby's Care" issued by health department to
replace 1919 publication; distributed to home of all new-
borns
Milwaukee placed on National Health Honor Roll
Convalescent homes and homes for the aged licensed and
supervised by the health department
1943 "The March of Health" weekly radio dramatizations produced by the health department
Milwaukee again placed on National Health Honor Roll;
contest discontinued
1944 Whooping cough and diphtheria immunization of all children started
Penicillin therapy for gonorrhea begun at city hospital

1945 School Hygiene Clinic opened
1946 "Rapid treatment center" started for syphilis patients at city hospital
1947 City-wide chest X-ray program
Tuberculosis Control Center established
1948 Benjamin Spock's *Baby and Child Care* replaced city pamphlet for distribution to families of newborn babies
Mobile child welfare clinic for outlying areas of the city
1949 Improved child immunization program used tri-immunol combined vaccine against whooping cough, diphtheria, and tetanus
Blue Cross hospital insurance began payment at city hospital
1950 Population: 637,392
Death Rate: 9.6
First woman, Margaret E. Hatfield, M.D., M.P.H., appointed as Deputy Commissioner of Health. She left in 1953 to join state Board of Health and in 1960 became health commissioner of Kenosha
Milwaukee Cancer Diagnostic Clinic, a cooperative venture with the health department, Marquette University Medical School, and the Milwaukee Chapter of the American Cancer Society, opened

Essay on the Sources

The research material upon which this book most heavily relies includes official city documents, numerous Milwaukee newspapers, and the manuscripts and papers of various voluntary and professional associations. The *Annual Reports* of the health department, issued regularly after 1868 with the exception of the decade of the 1880s, were most useful. They tended to be chatty accounts of the activities of health officials, mostly laudatory, but admitting of problems. They do not hold back on the limitations of common council appropriations, nor do health commissioners balk at criticizing their predecessors. The *Reports* contain the basic information about the growth of health services. Their vital statistics, growing in sophistication as the years progress, describe Milwaukee's health and disease patterns. The statistics are maddeningly inconsistent from year to year, making comparisons difficult and sometimes impossible. In some years agents used ward figures, in others they divided the city by age cohort, sex, or ethnicity. The health department issued *Monthly Reports* (statistical) during some years and after 1911 a monthly bulletin aimed at public education, the *Healthologist*, which includes much useful information about health department activities. The state Board of Health, formed in 1876, also issued *Annual Reports* and Bulletins which contained much useful city data. The Milwaukee common council *Proceedings* provided the municipal framework for health legislation, proposed, passed, and rejected. School Board *Proceedings*, Department of Public Works *Reports*, and other city documents rounded out the official word. Municipal reports are available in Milwaukee Health Department Library, the Wisconsin State Historical Society (Madison), and the Milwaukee Municipal

275

Reference Bureau. The health department has unpublished reports, vital statistics, and miscellaneous papers.

The city newspapers offered the best antidote to officialese. For following the daily and weekly developments of health problems, solutions, and dilemmas, and for a healthy skepticism about department rhetoric and claims, the newspapers cannot be matched as a source for understanding reform processes. Milwaukee offered presses of every political persuasion and ethnic dimension and, by using more than one at a time, the historian benefits from an expanded and ever-evolving source of information about public opinion, emerging theories, and everyday practices of yesterday's cities. The *Sentinel*, a Republican and Progressive daily, was most useful for complete city coverage and for middle-class reformist editorial ideology. It is WPA indexed through 1879 and recently expanded, due to the hard work and endurance of Herbert Rice at the Milwaukee Public Library Local History Room, through 1890. For contrast, the *Daily News* offered the working people's perspective and illustrated how it changed from criticism of reformer activity to support of it during the desperate 1890s. Other newspapers added local color and varying perspectives: the *Journal*, Democratic; the *Leader* (after 1912), Socialist; and, very important for Milwaukee, the ethnic presses (*Abendpost*, *Kuryer Polski*, *Vorwärts*, *Germania*, *Der Seebote*, *Herold*), which revealed immigrant perceptions of their rapidly changing lives in America.

Newspapers, of course, have their own problems. Just as historians read official documents with care, sifting and winnowing about the self-glorifications, so too must they approach newsprint critically. Editorial changes and economic exigencies can affect the news pages as well as the editorial pages. By comparing the same health-related events in multiple newspapers, I tried to get beyond the biases of this source. Another problem is superficiality of reporting, due either to limited space or to editors' perceptions of how much the public wants or needs to know. I found the reporting of the Kempster impeachment trial incomplete in all newspapers. The technical aspects of the medical discussion, alluded to, are not discussed anywhere to a degree necessary to understand the scientific debate. Since the actual transcripts have disappeared, this newspaper shortcoming became significant. In other cases superficiality resulted from political choice. Newspapers found it wise not to dwell on certain aspects of political debates deemed harmful to their editorial position. This limitation was avoided in large part by using multiple newspapers. Despite the problems with this source, this book relies heavily on information and perspectives learned from Milwaukee's daily reporters. When the limitations are realized, there is no better place to turn for the unfolding of city life in the nineteenth century.

The Wisconsin State Historical Society and the Milwaukee Area Research Centers house rich collections of archival material on Milwaukee's voluntary associations and their public health activities. Of most benefit were the papers of the City Club of Milwaukee, which offered thorough accounts of public health activities of a civic betterment association. Over one hundred boxes of well-organized material provided insight into the interrelationships between private and public agencies. My only regret was that the Municipal League did not leave similar papers to document the earlier period. I also found the papers of the Visiting Nurse Association, the Wisconsin Anti-Tuberculosis Association, and certain private individuals, such as Marion Ogden and Lizzie Black Kander, helpful. Many associations' annual reports, incomplete runs except for the Wisconsin Women's Club, can be found in the Wisconsin State Historical Society. The Milwaukee Academy of Medicine houses the proceedings, minutes, and correspondence of all the regular medical societies that preceded the Academy. These were invaluable for understanding the medical profession's views of public health and city health department activity.

Many scholars have done excellent work on the history of Milwaukee, and their efforts have helped mine enormously. The classic biography of the city is Bayard Still, *Milwaukee: The History of a City* (Madison: State Historical Society, 1948), and it maintains its place admirably in the face of newer scholarship. Gerd Korman, *Industrialization, Immigrants, and Americanizers: The View from Milwaukee 1866-1921* (Madison: State Historical Society, 1967) takes up the immigrant saga where Kathleen Neils Conzen, *Immigrant Milwaukee 1836-1860: Accommodation and Community in a Frontier City* (Cambridge: Harvard University Press, 1976) leaves off. Clay McShane, *Technology and Reform: Street Railways and the Growth of Milwaukee, 1887-1900* (Madison: State Historical Society, 1974) provides a straightforward account of a technologically and economically complicated part of Milwaukee's history. Three un-

277

published theses were of great benefit to my understanding of Milwaukee social and political patterns: Roger Roy Keeran, "Milwaukee Reformers in the Progressive Era: The City Club of Milwaukee 1908-1922," M.A. thesis, University of Wisconsin, 1969; Roger David Simon, "The Expansion of an Industrial City: Milwaukee 1880-1910," Ph.D. disseratation, University of Wisconsin, 1971; and Ann Shirley Waligorski, "Social Action and Women: The Experience of Lizzie Black Kander," M.A. thesis, University of Wisconsin, 1970. David P. Thelen's seminal work on Wisconsin progressivism, *The New Citizenship: Origins of Progressivism in Wisconsin 1885-1900* (Columbia, Mo.: University of Missouri Press 1972) early made me aware of the potential of studying public health reforms in this progressive state and especially in its major city.

Numerous historians of medicine have influenced my thinking and conceptualization of the development of governmental responsibility for public health. Most significant among them for this study include Charles E. Rosenberg, *The Cholera Years: The United States in 1832, 1849, and 1866* (Chicago: University of Chicago Press, 1962); Barbara Gutmann Rosenkrantz, *Public Health and the State: Changing Views in Massachusetts, 1842-1936* (Cambridge: Harvard University Press, 1972); Richard Harrison Shryock, *Medicine in America: Historical Essays* (Baltimore: Johns Hopkins Press, 1966); James H. Cassedy, *Demography in Early America: Beginnings of the Statistical Mind, 1600-1899* (Cambridge: Harvard University Press, 1969), and *Charles V. Chapin and the Public Health Movement* (Cambridge: Harvard University Press, 1962); Thomas McKeown, *The Modern Rise of Population* (New York: Academic Press, 1976); John B. Blake, *Public Health in the Town of Boston 1630-1822* (Cambridge: Harvard University Press, 1959); John Duffy, *A History of Public Health in New York City*, 2 volumes (New York: Russell Sage Foundation, 1968, 1974); and Nelson Blake, *Water for the Cities* (Syracuse: Syracuse University Press, 1956).

Another important source for this book was the periodical press. Local publications like the *Milwaukee Medical Times* and the *Wisconsin Medical Journal* and national ones such as the *American Journal of Public Health* and the *Journal of the American Medical Association* proved valuable for understanding medical opinion and for placing the local doctors in the larger national perspective.

One of the most helpful was the *Public Health: Papers and Reports of the American Public Health Association*, 1873-1911. Periodicals like *Survey*, *McClures*, *Harpers*, and *Charities* gave a muckraker or social-science perspective on health situations in other cities and also helped to place the Milwaukee experience in a broader context.

Index

Abraham Lincoln Settlement, health programs of, 197-198. *See also* Settlement houses; Kander, Lizzie Black

Adams v. Milwaukee, tuberculin case, 183

aldermen. *See* Common council

Allen, James M. (Milwaukee physician), thinks swill milk safe, 165-166

American Public Health Association: meetings of, 59, 131-132; advice on garbage disposal, 134-135, 143; sponsors health conservation contest, 214

antitoxin. *See* diphtheria, antitoxin

antivaccinationists: among physicians, 80, 246; among German immigrants, 80-83, 84-85, 88, 94-95, 99; among Polish immigrants, 85, 88, 94-95, 99

Archbishop, supports child welfare work, 220

Associated Charities, health work of, 192, 193-194

Babies Free Hospital, 218

Babies Fresh Air Pavilion, 195

Bading, Gerhard A.: elected mayor (1912), 21; appointed health commissioner, 48, 52; on smallpox, 118-119; garbage control efforts of, 151-152, 154; and tuberculin testing, 185, 186; opposes

Child Welfare Commission, 223-225

bakeries, 5

Beffel, John M. (Milwaukee physician): anti-tuberculosis work, 201; opposes Seidel's election, 217; supports child welfare, 217, 218

Bennett Law (1889), 19, 252

Bennett, W. S. (Milwaukee physician), health department analyst, 178

Benzenberg, G. H. (city engineer), on garbage problems, 142

Berger, Victor, Socialist politician, 21*n*, 110

Bernoski, Jennie, nurse at child welfare station, 224

Biggs, Hermann (New York health commissioner), compared to Milwaukeeans, 258

blacks, in Milwaukee, 13

Board of Health: established, 4, 42, 45-46; activities, 38-40, 42-75; temporary, 43-44, 78-79; structural modifications of, 46-47; converted to department of health, 47. *See also* Health department

Bohemian immigrants. *See* immigrants, Bohemian

Boorse, Lorenzo (Milwaukee physician): administers medical milk commission, 181; on staff of Ba-

281